Making Markets in Australian Agriculture

Patrick O'Keeffe

Making Markets in Australian Agriculture

Shifting Knowledge, Identities, Values, and the Emergence of Corporate Power

palgrave
macmillan

Patrick O'Keeffe
School of Global, Urban & Social Studies
Royal Melbourne Institute of Technology
Melbourne, VIC, Australia

ISBN 978-981-13-3518-1 ISBN 978-981-13-3519-8 (eBook)
https://doi.org/10.1007/978-981-13-3519-8

Library of Congress Control Number: 2018965205

Cover illustration: © rkankaro / Getty Images
Cover Design by Fatima Jamadar

This Palgrave Macmillan imprint is published by the registered company Springer Nature Singapore Pte Ltd.
The registered company address is: 152 Beach Road, #21-01/04 Gateway East, Singapore 189721, Singapore

For my aunty, Ria de Groot

Abstract

This work analyses the case study of wheat export market deregulation in Australia, which was implemented in 2008, ending 60 years of statutory marketing by the Australian Wheat Board (AWB). At the time, policy makers claimed that this policy change would empower individual growers, providing them with choice and freedom in wheat marketing. However, regional wheat markets have become concentrated and are increasingly controlled by a small number of transnational agribusiness firms.

I argue this shift should be viewed as part of the broader restructuring of Australian society and economy. In analysing how the deregulation of the wheat export market has been made possible, I examine the construction of knowledge, values, and identities, to conform to the rationality of liberalised markets. These constructions create a reality which makes the shift from the public to the private appear as a logical, common-sense solution to the challenges facing society. I use the case study of farming and, specifically, wheat export market deregulation to show how this shift has been made possible in this context. To make this reality operable, I show how governmental technologies such as audit, the entrepreneurial individual, cost-benefit analysis, performance objectives, econometric modelling, and the consumer were used to act upon society, to make the shift towards liberalisation of the wheat export market happen.

The construction of firms as passive efficiency maximisers is central to this shift. As a consequence, policy makers have either ignored or failed to recognise the capacity of firms to shape their external environments to create favourable operating conditions: a "business-friendly environment". Thus, the interests of firms have been portrayed by policy makers as essentially being the interests of the broader society. In the case of wheat export market deregulation, the liberalisation of this market has enabled transnational firms to expand their geographical footprint and extend their global value chains. On the other hand, farmers, whom policy makers claimed were the primary beneficiaries of wheat export market deregulation, contend with consolidated markets instead of choice, declining autonomy rather than individual freedom, and, in some cases, feelings of disempowerment and disenfranchisement.

Contents

List of Tables

CHAPTER 1

Making Markets: Agricultural Restructuring in Australia

Introduction

Why Does This Matter?

This book analyses discursive constructions of the market, consumers, producers, efficiency, competition, and productivity, and their application to the Australian wheat industry. Drawing on the work of Rose (1993, 1996), Miller and Rose (1990, 2008), Dean (1999), Higgins (2001a, b, 2002a, b, 2005), and Higgins and Lockie (2002) in particular, I analyse the discursive construction of these rationalities. I argue that policy makers have sought to construct the rationality of the market-based society as a reality, through the shaping of knowledge, values, and identities which accord with this reality.

Policy discourses construct knowledge as the result of quantitative analysis. In this sense, to quantify something is to know it. This construction of knowledge has significant implications for how we understand and interpret the world. In addition, it influences what matters in relation to policy, particularly policy which is portrayed by policy makers as evidence based. Furthermore, policy discourses have sought to shape what we value as citizens and as a society. For example, quality of life, as an overarching policy ambition, has been subtly replaced by living standards, which are portrayed as consumption-centric and can be appeased by firms in liberalised markets. In this regard, what we value, or what policy discourses suggest we should value, is achievable through the construction of an

P. O'Keeffe, *Making Markets in Australian Agriculture*,
https://doi.org/10.1007/978-981-13-3519-8_1

environment where firms' involvement in our lives is enhanced and markets are portrayed as the co-ordinator of these firms. Thus, a "business-friendly environment" is framed as a good development for society. Coupled with these shifts is the construction of identities as an important facet in the neoliberalisation of Australian society. For example, citizens are reconstructed as consumers. This claim is not novel. However, I argue that it is central to the reality of markets and firms, the rationality of neoliberalism.

The development of the "business-friendly environment", and the rise of the technology of the consumer, I suggest, has been central to the neoliberalisation of Australian agriculture. In addition, farmers' construction as individualistic and entrepreneurial business people shapes what it is to be a "good farmer". This creates a reality where increased private investment in Australian agriculture is portrayed as common sense. Furthermore, the construction of firms' identities is also important to this narrative. Firms are cast as being central to the betterment of society, as having an interest in society's welfare above all else. This portrayal is clearly misleading in the modern society, where the shareholder and their interests have been constructed as the primary concern of the firm (Morgan 2014; Fairbairn 2014; Lazonick 2012).

Whereas the deregulation of the wheat industry was framed as being in the interests of wheat farmers, this policy shift has created an environment where wheat farmers and their communities are disenfranchised and undermined. Instead, wheat export market deregulation has been introduced to encourage transnational agri-food firms to extend their participation in Australian agriculture. This policy environment has been designed to nurture the interests of these actors, in the flawed assumption that their increased power and control over Australian agriculture will benefit Australian society. In justifying policy change such as wheat export market deregulation as "in the national interest", policy makers obscure the effect of policies such as this upon people and communities. In the case of wheat export market deregulation, and related policies such as structural adjustment and Agriculture—Advancing Australia, the broad aim is to maximise industry efficiency and productivity, which is assumed to lead to greater contribution of agriculture and industries such as wheat to the national economy. However, these policies also reduce the importance of farmers and communities, distorting and externalising the negative social consequences of these policy shifts. Farmer exits are referred to vaguely as adjustments, which are the necessary consequences of these policy changes, which, in any case, only impact farmers of lesser competence. The social impact of these shifts is masked as short-term pain, without acknowledging the potential for farmer exits to undermine rural communities and social capital of rural towns. Furthermore, if there are benefits to these shifts which accrue to the nation, it is unclear which sections of the community receive these benefits and what these benefits might look like. It could be argued that this is measurable through a greater contribution of agriculture to Australia's gross domestic product (GDP). This argument presumes that consumers might ultimately benefit. However, this implies that the benefits to consumers (however unclear or unequally shared) outweigh the real and perceptible negative consequences for rural Australia—most particularly, farmers, their families, and communities reliant upon the farming population. The obvious beneficiaries of agricultural and related-economic restructuring are transnational firms and their shareholders, whose access to Australian markets allows them to expand their geographic footprint and therefore develop new markets and sources of supply, as well as institutional investors, whom the Australian government is encouraging to increase their investment in Australian farmland.

Beyond the Australian wheat industry and agriculture, this work contributes to research highlighting the flawed arguments and assumptions which have underpinned the restructuring of Australian society in recent decades. The processes and technologies which I highlight are not confined to the Australian wheat industry. Policy discourses have shaped perceptions of how we know, what we know, and what we value. These constructions have shaped policy to focus on increasing efficiency, competition, and individualism. The resulting "business-friendly environment", which is claimed to improve people's well-being, is characterised by the transference of power away from the public towards corporations.

My research highlights this process in relation to farming. This study focuses on farming, specifically the wheat industry, to show how technologies of agency and performance have been used to operationalise the reality of markets, consumers, and competition. However, the policy tools which have been used to make this shift happen—quantification, individualism, firms as efficiency maximisers, efficiency maximisation and living standards, consumer choice and freedom—are endemic throughout Australian society. These constructions are buttressed by the argument that increased allocative efficiency, increased private firms' participation in society, and maximised resource productivity are in the national interest. In this regard, the national interest is crystallised in constructed indicators such as the unemployment rate, GDP, and multifactor productivity growth. The process by which GDP is increased, the legitimacy of this indicator, or the distribution of the gains which improved economic performance GDP is claimed to reflect are portrayed by policy discourses as secondary concerns. Is this in the national interest? I suggest that it is not.

Throughout this work, I argue that policy discourses have changed what matters, in relation to Australian farming, the welfare of farming communities, and policy governing the farming industry in Australia. Policy discourses have changed what counts as legitimate, credible knowledge and, in the process, have changed what counts as legitimate, credible policy objectives. This changed reality has been operationalised by technologies of performance and agency, which have facilitated a shift from the public to the private. This shift is not necessarily a transference of power from the state to corporations, or a retraction of the state's powers. However, this shift is encapsulated by a shift in power from the public to the private—a creation of privatised spaces, where public spaces once existed. The market is constructed as a producer of information, as a decider of what is fair and what is not, and as an organiser of society. The key participants in markets, consumers and firms, are integral to this construction. Critically, policy makers have constructed a corporatised society, by assuming that in the relationship between these market participants, consumers have the power. This assumption is incorrect. Consequently, this reconstruction of society has shifted power from the public to private interests, which I argue is not in Australian society's interests or the interests of the nation.

Contribution of This Work

Throughout the 1980s, 1990s, and 2000s, successive Australian governments engineered extension programmes of economic reform, aimed at deregulating key industries, privatising state-owned assets and authorities, and rescinding trade barriers. This shift in policy focus was particularly pronounced in agricultural industries. As such, the restructuring of government intervention in agricultural industries has had a significant impact on farmers and their communities, which has been extensively documented by authors such as Lawrence (1987), Gray and Lawrence (2001a), Tonts and Jones (1997), McKenzie (1994), Cocklin and Dibden (2002), Botterill (2012), and Head, Atchison, and Gates (2012).

I aim to contribute to this literature by examining how such seismic shifts in policy were made possible, and in whose interests. In particular, I focus my analysis upon the discursive dimension of power relations, expressed through the economic rationalist programmes enacted by successive Australian governments. I analyse the Australian wheat industry as a case study. The Australian wheat export market was one of the last remaining markets to be deregulated. In 2008, the statutory marketing of wheat was ended by the Australian government, which allowed private grain traders to export Australian bulk wheat for the first time since 1948 (Botterill 2012). The deregulated wheat market was predicted to feature multiple grain traders competing amongst themselves for growers' wheat, empowering growers and enabling them to exercise choice over whom to sell their wheat to (Commonwealth of Australia 2008a, b, 2012). However, as I show in Chap. 6, this scenario has not eventuated. Regional markets have become concentrated, whereas growers have expressed concern that the promised benefits of deregulation have not materialised (O'Keeffe and Neave 2017; Baker 2016, 2018).

The deregulation of the Australian wheat export market is underrepresented in geography and sociology research in particular, with literature on this topic dominated by scholars from the field of agriculture and resource economics, whose research has primarily centred on analysing and predicting the changes to wheat prices and supply chain costs resulting from this shift (Mugera, Curwen, and White 2016; McCorriston and McLaren 2007; Irving, Arney, and Linder 2000; Allen Consulting 2000; Royal Commission into Grain Handling, Storage and Transport 1988). As such, scholarly and government-produced literature, which emphasises the impact of market structure upon wheat prices and costs, has narrowed "what matters", in relation to this policy area. Furthermore, I suggest in my research, knowledge, values, and identities around the wheat industry and farming have been constructed to value market-driven data as credible and objectives such as maximising wheat prices as the only legitimate industry policy aim.

This research draws on research into governmentality, political economy, rural sociology, sociology of quantification, and financialisation. In utilising these diverse bodies of work, I suggest that the issue is much more complex than simply assessing changes to market-driven data such as prices and costs. Thus, I focus on how policy discourses have sought to construct market-produced data as legitimate, how the construct of the good farmer has been shaped to promote self-reliance and entrepreneurialism as a good farming attitude, and how policy values have been shaped to prioritise efficiency and marginalise equity. These discursive constructions are all integral to making deregulation possible. In analysing the neoliberalisation of Australian agriculture, through focusing on the reconstruction of knowledge, identities, and policy values while underpinning Australian society, I draw upon Higgins and Lockie's (2002, p. 420) conceptualisation of neoliberalism, which they argue:

> [...] needs to be seen as more than just the application of a political philosophy, ideology or the evolution of a new state form. Rather, it comprises an assemblage of rationalities and technologies of governing that seek to govern in an *advanced liberal way* [...] rather than portending an abandonment by the state of the will to govern, advanced liberal rationalities of government seek to transform critiques of the welfare state into strategies for governing conduct in more effective ways.

Whereas this work analyses what is an apparently simple shift from state intervention (through statutory wheat marketing) towards liberalised markets, I use this research to show how technologies of government have been employed, in the past three decades in particular, to make this shift appear to be a logical and common-sense response to the deceptively simple question of which market structure is best equipped to maximise growers' returns.

In the remainder of this chapter, I provide background to the issue of agricultural restructuring, both in Australia and in other prominent nations in agriculture. I outline agricultural restructuring internationally in recent decades, focusing on the structural causes of farmer exits, and the impact that this has on rural communities and farmers. Then, I analyse changing Australian agricultural policy, before focusing on changing wheat export market policy throughout the twentieth century, through to the liberalisation of this market in the mid-2000s. First, I turn to governmentality, which offers an important lens for understanding how wheat export market deregulation came to be portrayed as a logical, common-sense policy solution and how the social consequences of this shift were effectively erased from the policy-making consciousness.

Applying Governmentality in the Context of Australian Agricultural Deregulation

The deregulation of Australian wheat marketing is relatively unexplored by geographers or rural sociologists, who are otherwise well positioned to analyse these policy changes. While there has been a wealth of literature from Australian rural sociologists and geographers exploring the effects of deregulation upon rural communities and farmers in the past few decades (Argent 2005; Vanclay 2003; Gray and Lawrence 2001a, b; Tonts and Jones 1997; McKenzie 1994; Lawrence 1987), wheat export market deregulation is a substantial policy change which, until recently, has been primarily studied by agricultural economists (Mugera et al. 2016; Williams 2012; McCorriston and MacLaren 2007). Sociologists provide glimpses of the effect of wheat export market deregulation. However, this is often provided within the context of broader studies that explore social capital (Talbot and Walker 2007) or farmers' experience of climate change (Head, Atchison, Gates, and Muir 2011). These studies point to issues arising from deregulation without, however, fully allowing deregulation to take centre stage and thereby opening a problem space that demands further attention.

In this work, I draw upon governmentality as a lens for analysing Australian agricultural restructuring and, specifically, the liberalisation of the wheat export market. Governmentality has been widely used in an Australian context, to understand economic restructuring in relation to rural, regional, and agricultural policy change (Dufty-Jones 2015; Argent 2011; Gibson, Dufty, Phillips, and Smith 2008; Cheshire and Lawrence 2005; Larner 2000; Higgins and Lockie 2002). Governmentality research analyses the "conduct of conduct", where society is governed at a distance, through governmental technologies which establish legitimate ways of knowing, being, and acting, according to political aims and ideologies (Lockie and Higgins 2007; Argent 2005; Dean 1998, 1999; Rose 1996). This approach can be used to understand roll-out neoliberalism, to identify the power which is exercised on people and communities, despite the claimed freedoms, choices, and empowerment promised by discourses of marketisation and the individual.

For example, the deregulation of the wheat export market could be interpreted as a simple transition from government regulation of markets to market liberalisation. This depiction of wheat market deregulation, which is reflected in the agricultural economics literature on this topic, implies that the neoliberalisation of agriculture is understood as the withdrawal of state intervention in markets and the reduction of state assistance for farmers. This process is characterised by Peck and Tickell (2002) as "roll-back" neoliberalism, most notably employed in the 1980s by the British and US governments of Thatcher and Reagan, though also by the Hawke and Keating governments of the late-1980s and early-1990s in

Australia. However, as Peck and Tickell (2002, p. 384) have argued, in the time since:

> [...] the agenda has gradually moved from one preoccupied with the active destruction and discreditation of Keynesian-welfarist and social-collectivist institutions (broadly defined) to one focused on the purposeful construction and consolidation of neoliberalised state forms, modes of governance, and regulatory relations.

According to Peck and Tickell (2002), this second wave of neoliberalism resembles a programme of reregulation, or "roll-out neoliberalism". Whereas "roll-back" neoliberalism suggests the withdrawal of government control of social and economic life, "roll-out" neoliberalism has come to resemble "a range of rationalities and techniques that seek to govern without governing society, to govern through regulated choices made by discrete and autonomous actors" (Rose 1996, p. 328). These techniques, operationalised by neoliberalism, are described by Rose (1993, pp. 294–295) as including "monetarisation, marketisation, enhancement of the powers of the consumer, financial accountability and audit". These technologies derive their power from their apparent disinterest and harmlessness (Miller and Rose 1990). Although they imply a level of distance from the state, governmentality theorists argue these technologies are instead used by the state to shape the conduct, the behaviour, the attitudes, and values of its citizens.

This research draws on governmentality as a lens for understanding the construction of individual identities, particularly farming identities, knowledge, and values, to make wheat market deregulation appear as a logical shift, according to what farmers represent, what is known about the wheat industry, and what matters in relation to Australian farming. I suggest that these constructions helped to create a reality of farming in Australia, which was then used to shape how farming could be understood and acted on by policy makers. In doing so, I utilise Dean's (1999, pp. 167–169) analysis of technologies of agency and performance, to understand how these constructions have been carefully curated in policy discourses around competition, economic policy, and farming. Similarly, Higgins' work (2001a, b, 2002a, b) provides an excellent example of how governmentality has been applied to understand neoliberalisation in an Australian farm context.

This leads to the role of discourse in shaping values, identities, and knowledge which accorded with the rationality of markets and which understood markets as the producers and communicators of value. In focusing on the work of Miller and Rose (1990, 2008), Rose (1993, 1996), Dean (1999), Higgins (2002a, b), and others, I aim to build upon previous research which critically analyses regional policy through exploring discursive formations (Lockie and Higgins 2007; Cheshire and Lawrence 2005; Liepins and Bradshaw 1999; Liepins 1996; Sharp and Richardson 2001). In undertaking this research, I have collected and analysed policy documents produced in Australia since the early 1970s, which share a focus on competition policy, economic policy, and agricultural policy. I use this research to analyse the construction of policy truths such as efficiency, productivity, the rational individual, and "the market", focusing on how these truths have been used to govern Australian agricultural policy change since the 1980s. Thus, this research has a genealogical focus, as I analyse documents produced across four and a half decades, to identify the subtle shifts which have occurred throughout this period and which help us to understand how the present can be understood. I now provide background to this case study, by describing shifting agricultural policy in Australia since the Second World War, in relation to policy change in countries such as Canada, the United States, and New Zealand.

A Brief Outline of Recent Agricultural Policy Change

The problematisation of the state is particularly evident in discourses around Australian agricultural and regional policy, which has contributed to substantial policy change in these areas in recent decades. However, this is not unique to Australia. Following the Second World War, in Western countries such as the United States, Canada, Australia, and New Zealand, agricultural policy was characterised by high levels of government intervention. This policy environment supported a collectivist, cooperative approach to agricultural organisation, evident in policies such as statutory marketing of commodities, government underwriting of commodities prices, and the use of tariffs. Policy makers focused on ensuring the security and stability of farmers, reducing their exposure to risk and uncertainty (Ryan 1984). This approach changed across the 1970s and 1980s. Global commodities' markets weakened, placing considerable pressure on this approach to agriculture (Cockfield and Botterill 2007; Coleman and Skogstad 1995). This shift threatened the viability of smaller farms in the United States, Australia, and Canada, and in smaller producing countries such as Austria and Finland (Dixon and Hapke 2003; Pietola, Vare, and Lansink 2003; Lobao and Meyer 2001; Tonts 1999; Weiss 1999). In Australia and the United States in particular, this contributed to economic crises developing within agriculture, threatening the long-term existence of farming communities which had traditionally been supported by the presence of family farmers (Dixon and Hapke 2003). Consequently, national governments had to decide whether to continue to protect and subsidise agriculture, or to retreat from this position. In countries such as Canada, the United States, New Zealand, Australia, and South Africa, the government response has been to liberalise agricultural markets and to rescind policies which had been implemented to protect and stabilise agricultural industries (Bradshaw 2004; Liepins 2000; Mather and Greenberg 2003; Skogstad 1998).

In New Zealand, successive governments have restructured agriculture, exposing industry and farmers to global competition (Woods 2006; Liepins 2000; Le Heron and Roche 1999). Le Heron and Roche (1999, p. 204) highlight the reduced government support for agriculture in New Zealand, which is emphasised by a reduction in subsidies from 34% of farm receipts in 1984, to "almost zero" in 1995. However, as Le Heron and Roche (1999, p. 204) argue, the removal of policies and institutions aimed at stabilising and protecting New Zealand agriculture did not result in an absence of government intervention or regulation. Instead, as Le Heron and Roche (1999, p. 204) suggest, "agriculture has been re-regulated in a potentially more complex way, through greater market disciplines, with an emphasis on certain kinds of political intervention, and is not simply a rolling back of the state." This leads to an important point—removing forms of political intervention, such as farm subsidies, might represent a shift away from state control in agriculture. However, as I suggest in this work, this is not necessarily the case.

In the United States, pressure from farm lobby groups, coupled with the esteemed position of the family farm in the American psyche, contributed to the development of the 1985 Farm Security Act 1985, which Dixon and Hapke (2003, p. 150) describe as "the most expensive outlay on agriculture in US history". This policy reinforced state support for farmers. However, the 1996 Federal Agricultural and Improvement Reform (FAIR) Act represented a drastic shift towards agricultural market liberalisation (Dixon and Hapke 2003; Brasier 2002; Skogstad 1998). As explained by Skogstad (1998, p. 463), this policy created a significant break from previous approaches to agriculture in the United States:

> Long-standing interventionist policy instruments were eliminated; new principles governing relations between the US government and its grains sector were installed; and productivist, export-oriented goals became uppermost.

Dixon and Hapke (2003, p. 142) cite then President Bill Clinton, who claimed that "At long last, farmers will be free to plant for the markets, not for government programs." However, as Brasier (2002, p. 239) highlights, this policy shift was characterised by competing discourses, which on the one hand reiterated the government's commitment to supporting farmers and shielding them from market volatility, and on the other hand professed an intention to reduce government intervention in agricultural markets, thus enabling farmers to exercise their freedoms and take advantage of changing market conditions.

Skogstad (1998) contrasts the Clinton administration's 1996 FAIR Act with the 1992 MacSharry reforms to the European Union's Common Agricultural Policy (CAP), which Skogstad claims contributed to ongoing state support for agriculture in Europe. In addressing the question of why the approach in Europe differed from that of the United States, Skogstad (1998, p. 467) suggests that the agricultural exceptionalism which underpinned the European Union's state-assisted approach to agriculture had a "better fit with the larger European ideational framework", while also showing the capacity to shift, taking on "additional, new goals which gained it adherents beyond the farm policy community". As Skogstad (1998, p. 467) states, "This did not happen to the same extent in the USA, where by the late 1990s the underlying productivist, export-oriented logic of US agricultural policy was at odds with a model of state assistance."

By contrast, European agricultural policy is underpinned by a multifunctional approach, which assumes that "without state assistance, large numbers of family farmers are vulnerable to economic pressure," potentially resulting in farmers abandoning their land (Dibden, Potter, and Cocklin 2009, p. 303). This pressure is constructed as having potentially severe social and environmental consequences. Thus, policies maintaining trade barriers and assistance to farmers are constructed as an essential measure to protect the European countryside from global markets (Dibden, Potter, and Cocklin 2009; Potter and Tilzey 2005, 2007). As Dibden Potter and Cocklin (2009, p. 302) argue, "the public good role of farming is key to this understanding, the argument being that farming (particularly in marginal areas) needs to continue if farmlinked biodiversity and the appearance and amenity of rural landscapes are to be maintained."

Potter and Tilzey (2005, 2007) and Dibden, Potter, and Cocklin (2009) have noted that this approach to agricultural policy contrasts with the increasingly neoliberal policies of the European Union in other areas, such as financial and labour market deregulation. However, this ongoing support for agricultural assistance for farmers is attributed to the support for multifunctionality in agriculture, which is grounded in a "social welfare justification for state assistance" (Potter and Tilzey 2005, p. 590). This approach has influenced policy makers, who Potter and Tilzey (2005, p. 590) argue have "gradually acknowledged the need to diversify the income base of family farms by capitalising on agriculture's ancillary functions such as biodiversity, landscape and cultural heritage". The multifunctional approach to agriculture, though weak in Australia (Bjørkhaug and Richards 2008; Argent 2002), is used in countries such as France, Germany, and Norway, with the latter using government subsidies to employ the "language and action of a multifunctional agriculture into its agricultural mode of operation" (Bjørkhaug and Richards 2008, p. 108). Bjørkhaug and Richards (2008) and Dibden, Potter, and Cocklin (2009) highlight the role of farming organisations in supporting multifunctionality and maintaining state assistance for farmers. This approach differs significantly from the Australian example that I turn to now.

Deregulation in Australian Agricultural and Regional Policy

In this section, I briefly highlight some of the key changes which contributed to this shift, before then drawing on the effect that this has had upon farmers and their communities. In recent decades, successive Australian governments have deregulated agricultural industries by removing state assistance, liberalising markets, and privatising government-owned assets and authorities (Tonts and Haslam-McKenzie 2005; Alston 2004; Vanclay 2003; Cocklin and Dibden 2002; Vanclay and Lawrence 1993). The development of the Industries Assistance Commission (IAC) in 1974 was a critical aspect of this shift (Edwards 1987). The IAC was created by the Whitlam government to assist the state to maximise productive use of resources, and to develop policy which would not harm domestic consumers (Warhurst 1982; Industries Assistance Act 1973). This change in policy focus, and the subsequent withdrawal of state support for agriculture, fundamentally altered the policy environment in Australia (Edwards 1987; Warhurst 1982). Examples of this change include the 25% across-the-board drop in tariffs in 1973, which signalled the intention of the Whitlam government to shift policy emphasis away from protectionism, towards market-based approaches to economic organisation (Anderson, Lloyd and McLaren 2007).

The 1971 Rural Reconstruction Scheme was the first attempt of the postwar Australian governments to facilitate the removal of assistance for farmers whose farms were deemed to be "unviable" and to reallocate resources towards more profitable, efficient farmers (Cockfield and Botterill 2006, 2007; Higgins 2001a, b; Musgrave 1979). Musgrave (1990) provides an example of the positive construction of adjustment through farm reorganisation and resource reallocation:

> In the final analysis, the process of labour shedding and farm reorganisation should be regarded as desirable in terms of economic efficiency. Presumably those who leave the industry have good reason and are better off for doing so. Similarly, those staying in the industry should also benefit. However, in an ongoing adjustment process there is no final analysis and the process is not without its costs either to individuals or the nation as a whole. Evidence of the cost to individuals is seen in terms of chronic and ephemeral poverty among rural people which, in turn, is a cost to the nation if it finds this form of poverty abhorrent. In addition, there may be other costs to society, springing from inefficient resource use due to lags in the adjustment process [...] The conclusion would seem to follow that, if there is a problem requiring government intervention, it would be because adjustment is not occurring. In this respect the persistence of low farm income problems could be an indication that adjustment is not occurring fast enough.

In this regard, the "problem" of government intervention is framed, not as one of preventing inefficient farmers from exiting the industry but as one of concern that this process may not be occurring quickly enough. This perspective gained traction across the 1980s with the election of the Hawke government in 1983, which further reconceptualised the relationship between farmers and the state. Rather than providing passive assistance to farmers, the Hawke government perceived its role as creating an economic environment which enabled the most efficient farmers to be successful (Hawke 1986; Kerin 1986). Questions around agricultural policy became focused on how to maximise efficiency and productivity of resource use. Subsequently, the value of agricultural industries and the contribution of farmers to Australian society were judged according to these ambitions. Similar processes of deregulation have been implemented across other agricultural industries in Australia, including barley (Botterill 2012), dairy (Margetts 2007; Cocklin and Dibden 2002; Davidson 2001, 2002), beef (Pritchard 2006), wine grapes (Pritchard 1999), egg (Alston 1986), and wool (Vanclay 2003). Highlighting its influence over policy, the Industry Commission (1991, p. 23) claimed to have:

> [...] examined the effects of statutory marketing arrangements in recent inquiries into apples and pears (1990), the food processing and beverages industries (1989), the dried vine fruits industry (1989), the wheat industry (1988), the fresh fruit and fruit products industries (1988), the tobacco growing and manufacturing industries (1987) and the rice industry (1987).

These examinations typically reached similar conclusions—namely, that statutory marketing, particularly involving compulsory acquisition, undermined efficient resource use (Industry Commission 1991).

Neoliberalisation, Farm Exits, and Rural Communities

The shift away from interventionist approaches to agriculture in the latter part of the twentieth century contributed to declining farming populations throughout the world. It should be noted that technological development, which allowed farmers to cover greater expanses of farmland with fewer employees, is a major cause of this decline. Nevertheless, a major change, albeit partly facilitated by technological development, is farm consolidation. As highlighted above by Musgrave, this process has been accelerated in Australia, by governments' intent on extracting maximum productivity from the nation's resources, through encouraging allocative efficiency. Thus, policies such as the 1971 Rural Reconstruction Scheme, the Structural Adjustment packages, the Agriculture—Advancing Australia package, and Wheat Export Market Deregulation have each sought to facilitate the exits of the least efficient farmers from the industry, in the belief that this would liberate the resources held by these farmers, to be used more productively by farmers of larger scale, with greater access to technology.

As a result of these processes, in Australia, throughout Western Europe and European countries such as Finland and Austria, the United States, and Canada, the numbers of farms dropped significantly. The number of farms in the United States declined from 6.5 million in 1920, to 2 million in 1992, with the proportion of the US population living on farms dropping from 30.1% to less than 2% in the corresponding period (Wittmaack 2006, p. 45). In Austria, the number of farms declined by 30.8% between 1960 and 1993 (Weiss 1999), whereas in Finland, policy change seeking to facilitate the early retirements of farmers caused the number of farms to decrease by 19% between 1995 and 2000 (Pietola et al. 2003). The decline in farm numbers reflects the increasingly concentrated nature of farming structures across a range of different countries and regions (Lobao and Stofferahn 2008; Breustedt and Glauben 2007; Weiss 1999). This shift is attributed to the increased capacity of technology to allow for greater farm sizes (Pietola et al. 2003) and a greater focus on industrialised farming, in place of the once revered notion of the family farm in countries such as the United States and Australia, as well as reduced financial support for farmers from the state (Lobao and Stofferahn 2008; Dixon and Hapke 2003; Brasier 2002; Lobao and Meyer 2001). This reflects shifting constructions of what contribution farming makes to national prosperity, shifting conceptualisations of the importance of farming to the identity of Westernised nations and, potentially, the declining importance of rural industries and communities in policy making. In turn, restructuring in the farm sector and the "displacement of farmers from farming" in these countries was viewed by neoclassical economists as "an indicator of the system's success" (Lobao and Meyer 2001, p. 110).

Within Australia, a combination of these factors has contributed to a similar decline in the number of farmers managing their own properties (Alston 2004, 2012; Tonts 1999). Between 1982 and 2011, Australia's farm population dropped from 263,200 to 157,000 (Australian Bureau of Statistics 2012). The exit of farmers from their land, and from the farming industry, affects farmers, their families, and their communities, in many ways. The increased concentration of farming, evident in Australia and observed in the United States, Europe, and Canada, for example, is associated with greater inequality within farming (Lobao and Stofferahn 2008; Wittmaack 2006). This shift has caused a polarisation within farming, as the middle is essentially hollowed out, with larger farms increasing in size and scale, and smaller farms becoming increasingly marginalised (Lobao and Stofferahn 2008; Breustedt and Glauben 2007; Lobao and Meyer 2001; Weiss 1999).

Research conducted in the United States, Europe, and Australia has found that high rates of farm exits have a negative, cyclical effect on rural communities (Lobao and Stofferahn 2008; Breustedt and Glauben 2007; McKenzie 1994). As mentioned by Breustedt and Glauben (2007, p. 115), in a western European context:

> The declining number of farms not only has consequences for the agricultural sector but also for rural areas as a whole. The loss of farms might lead to a depopulation of the countryside which in turn affects the demand of services and infrastructure of local communities.

The considerable decline in Australian farming populations significantly undermined many rural communities in Australia, particularly in the late Twentieth Century when this trend began to accelerate (Lockie and Bourke 2001; Lawrence 1987). Rather than remaining in their communities, farming families exiting their properties typically migrate to coastal regions, regional centres, or cities (Smith and Pritchard 2012; Alston 2004; Tonts 1999). As a consequence, the increased rates of farmer exits have undermined rural populations and, in some cases, threatened the viability of rural communities (Talbot and Walker 2007; Tonts and Jones 1997; McKenzie 1994). As McManus et al. (2012, p. 27) and Pritchard et al. (2012) have noted, farmers perform a key role in contributing to the resilience of rural communities, by maintaining local economies through employment, either directly in farming or ancillary businesses and community life, and through the maintenance of social spaces such as halls, as well as sporting clubs. Thus, declining farming populations contribute to a cycle of population out-migration and subsequently, the withdrawal of health, education, transport, and communications services, and therefore, local employment opportunities (Tonts and Jones 1997; McKenzie 1994; Stilwell 1992). As McKenzie (1994, p. 256) states, "There are also significant psychological impacts, with many seeing the closure of such central communities' services as signalling the 'death' of a town." The consequences of farmer exits upon the well-being of farming families and their communities are clear, and illustrate the limitations of policy making that has attempted to construct agriculture principally in economic terms. The effect of farmer exits, and increasingly concentrated and industrialised farming, is summarised by Lobao and Stofferahn (2008, p. 229), who state that this shift contributes to increased economic inequality, greater reliance upon government services, diminished social well-being, as well as negative environmental consequences.

Impacts of Farm Exits Upon Farmers' Identity, Well-Being, and Relationship with the Land

Economic restructuring in Australian agriculture has been implemented by national governments which have intended to enhance allocative efficiency within the sector. This has distorted the effect of farm exits upon farmers themselves. However, this effect is considerable, and it is due to the very nature of farming itself. As Lobao and Meyer (2001, p. 118) explain:

> The farm enterprise is inextricably connected to the household, so that production changes become reflected in work roles, hardship, stress, and resilience. As farms decline in number and grow in size, their effects reverberate across communities.

For many farmers, their identity, well-being, and sense of belonging are particularly grounded in their rural, farming environment (Kuehne 2012; Wiseman and Whiteford 2009; Wythes and Lyons 2006; Pretty, Chipauer, and Bramston 2003). This relationship has been studied by Cheshire, Meurk, and Woods (2013) and Gosling and Williams (2010) in an Australian context, in the United Kingdom (Raymond et al. 2016), and the United States (Walker and Ryan 2008). This research often focuses on the relationship of farmers to the land, and the influence that this relationship has upon their conservations behaviours (Raymond et al. 2016; Gosling and Williams 2010; Walker and Ryan 2008).

However, the severing of this relationship, through unplanned exits from farming, is particularly significant (Kuehne 2012). As Kuehne (2012, pp. 2–3) explains in his auto-ethnographical article on his experience of having to sell the family farm, this connection to past and future generations becomes entrenched within the identity of the farmer, as "farmers are not simply representing their own identities, but also the identities of their families stretching from the past to the future." Thus, exiting the family farm can not only result in the severance of these connections to the land, to family history, and the loss of income, it can also substantially impact farmers' sense of identity, sense of belonging, and, subsequently, mental health and well-being. As Kuehne (2012, p. 5) describes, beyond the financial transaction of selling the farm and the subsequent loss of livelihood:

> [...] there were even deeper and more profound changes than these. It changed who I associated with, my sense of importance, my place in the community and my relationship with that particular piece of land. The sale of my farm invoked deep feelings of loss. This was caused by more than the loss of the farm; it was loss of certainty, loss of purpose and loss of identity.

Furthermore, some farmers internalise the decision to sell their farm as a sign of personal failure (Kuehne 2012). This is particularly significant, as I argue later in this book, because policy discourses have portrayed farmer exits as the result of the poor farm management practices of the farmers who are leaving the land. This framing constructs farmer exits as a result of individual inadequacies, and as Kuehne (2012) discusses, farmers in turn have accepted individual responsibility for this failure to succeed in the occupation which is intricately connected to their sense of identity. As a result of this complex relationship between farmers, their land, family history, and identity, the considerable number of farmer exits in Australia throughout recent decades has resulted in significantly higher rates of anxiety, depression, and suicide among farmers, than is evident in the broader Australian population (Bryant and Garnham 2014; Alston 2004, 2012; Dean and Stain 2010; Wiseman and Whiteford 2009; Wythes and Lyons 2006; Fraser et al. 2005; Caldwell, Dear, and Jorm 2004).

The Australian Response to Declining Rural Fortunes: Competitive Productivism and Self-Reliance

The decline of Australian farming populations and subsequent effects of this out-migration from rural towns are closely associated with the retraction of state assistance for farming. As Dibden, Potter, and Cocklin (2009, p. 302) observe, Australian farmers experience the second lowest level of support among all Organisation for Economic Co-operation and Development (OECD) countries. According to Dibden, Potter, and Cocklin (2009, p. 302), farmers are:

> [...] encouraged to adapt to liberalised trade through development of more productive farming systems, i.e., through intensive farming practices, farm

consolidation, displacement of smaller, less 'efficient' farmers and developing better business management capacities.

This process is described by Cocklin and Dibden (2005) and Dibden, Potter, and Cocklin (2009) as a form of competitive productivism that has driven Australian agricultural policy. On the one hand, this is based upon the notion of allocative efficiency as a means of enhancing the competitiveness of Australian agricultural industries. On the other hand, this is partly claimed to be a response to rural decline, and the increased pressure experienced by many farmers to remain viable.

The competitive productivism that characterises agriculture in Australia has also been connected to damaging environmental consequences that result from intensive farming methods and the use of fertilisers, herbicides, and pesticides (Lawrence, Richards, and Lyons 2013; Pritchard, Burch, and Lawrence 2007). The pressure to maximise productivity to cope in an economic environment with few avenues for financial assistance compels farmers to use potentially damaging farming methods (O'Keeffe 2016a; Lawrence, Richards, and Lyons 2013). This is not to say that Australian farmers are not interested in minimising their environmental impact, or that all farmers are responsible for implementing damaging environmental practices. Far from it. The success of volunteer-based farm programmes such as Landcare is evidence of farmer engagement with environmental practices, as is the growing proportion of farmers employing organic farming methods (Tennent and Lockie 2013). However, Australian agricultural policy discourses have, for a number of decades, prioritised and rewarded productivity. As I argue in Chap. 4, productivity has been constructed as a policy truth, and productivist farming methods, such as the application of fertilisers, herbicides, and pesticides, concentration of farm properties, and increased specialisation, have come to represent smart farming (Argent 2002). Furthermore, environmental protections, such as policies which aim to prevent land clearing and mature tree removal, have been constructed as burdensome regulation, or "green tape" (Productivity Commission 2017).

By contrast, in Europe, farmers are provided with state support to modify their farming practices, in the interests of meeting environmental goals (Kleijn and Sutherland 2003). As described by Kleijn and Sutherland (2003, p. 947), agri-environment schemes are used for "reducing nutrient and pesticide emissions, protecting biodiversity, restoring landscapes and preventing rural de-population". As such, many of the programmes delivered for farmers are tied to specific social and environmental aims, which, in contrast with Australian agricultural policy, are considered by policy makers to have value.

Australian state and federal governments have responded to the social and environmental problems experienced throughout rural and regional Australia, by developing regional policy encouraging communities to become more self-reliant (Beer et al. 2005; Cheshire and Lawrence 2005; Tonts and Haslam-McKenzie 2005; Gray and Lawrence 2001b; Herbert-Cheshire 2000). Responsibility—though, crucially, not power—has been devolved from governments to local communities, who have been encouraged to find solutions to their own problems (Cheshire and Lawrence 2005; Lockie 1999). Regional development programmes such as the Victorian government's "Putting Locals First Program" have been developed according to carefully constructed funding criteria, closely aligning with the government's economic goals (O'Keeffe 2014). This approach partly absolves governments of responsibility for the ongoing resilience of rural communities, while ensuring that community responses to their challenges are tightly controlled and politically acceptable.

Similarly, agricultural policy in recent decades has focused on promoting individualism, competition, self-help, and self-reliance (Cheshire and Lawrence 2005; Gray and Lawrence 2001b; Higgins 2001a). I extend this analysis by exploring the conceptualisation of farmers as business-minded individuals who are centred exclusively in maximising their returns. This construction differs substantially from research analysing the importance of farmers' connection to their land, their family histories in farming, and the relationship between farming, identity, and well-being (McManus et al. 2012; Kuehne 2012; Fraser et al. 2005; Smailes 2000; Gray and Lawrence 1996). Policy discourses have constructed farming in economic terms, largely externalising these important social aspects of farming. As I will argue, this is an essential discursive shift that has helped to make wheat industry deregulation possible.

Focus on Shifting Wheat Marketing Policy: From Stability and Security to Individualism and Competition

Australian wheat industry policy has changed markedly throughout the past 100 years. Popular perceptions of the state's role in regulating farming, the trustworthiness of grain traders, and the role of the farmer in relation to the broader Australian society have changed on numerous occasions throughout this time, shaping wheat industry policy. In addition, global crises such as war and economic depression have often been the catalyst for policy change.

From 1948, with the federal government's introduction of the first Wheat Stabilisation Plan, the Australian Wheat Board (AWB) became the sole exporter of Australian wheat. For growers, this process was relatively simple (Pritchard 1998). Wheat marketing was conducted entirely by the AWB, which pooled the wheat it received from growers, according to classification. This allowed the AWB to amass large quantities of wheat to sell to overseas buyers. The scale of these transactions was held up as a key advantage of statutory marketing, which was argued to provide growers with collective bargaining power in world markets. Growers were essentially provided with average prices through this pooling system, as net returns to growers were calculated based on the value of wheat sales and the costs associated with these transactions, including freight charges and marketing costs. Until 1989, this process was supported by government guarantees on wheat prices, articulated in the Wheat Stabilisation Plans. According to Ryan (1984, p. 117), the first five plans set price guarantees according to the cost of production, whereas the Sixth Plan (1974–1975) shifted this approach to reflect export prices. The removal of guaranteed wheat prices exposed growers to global wheat markets, shifting the policy focus from stability to market efficiency (Pritchard 1998; Ryan 1984). In this climate, the AWB was increasingly scrutinised, with attention focused on its capacity to deliver premium prices for growers.

Furthermore, as highlighted in Fig. 1.1, throughout the early-to-mid-1980s, global wheat prices declined substantially in the midst of a "trade war" involving the United States and European Commission (Coleman and Skogstad 1995, p. 244). This price decline threatened the viability of farmers, causing the Canadian and Australian governments to analyse their own approaches to wheat industry assistance and marketing. At the time, the wheat industries in these countries were dominated by statutory marketing authorities, the AWB, and the Canadian Wheat Board, respectively, which were substantial participants in the global wheat trade (Boaitey 2013). Whereas the Canadian and Australian governments each sought to motivate trade liberalisation, the Canadian government also sought to provide a level of protection for growers (Martin and Clapp 2015; Magnan 2015; Coleman and Skogstad 1995). As highlighted by Coleman and Skogstad (1995, p. 257), the Canadian government

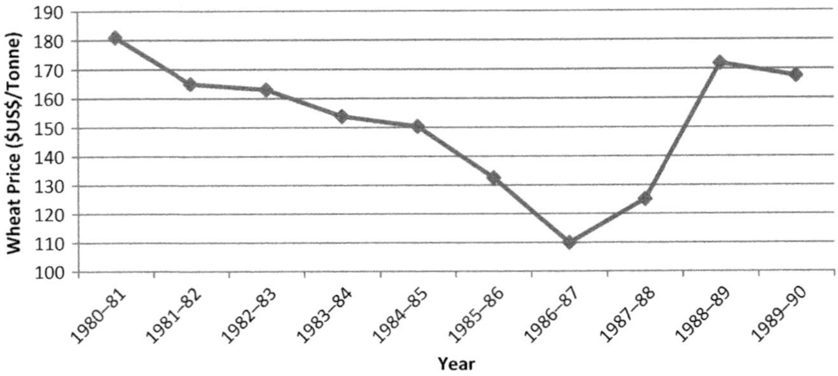

Fig. 1.1 Australian wheat price (US$/Tonne), from 1980–1981 to 1989–1990. (Source: Australian Bureau of Agricultural and Resource Economics and Sciences 2012)

responded to the pressure upon Canadian farmers by providing "ad hoc payouts to western farmers to shield them from the trade war". Coleman and Skogstad (1995, p. 257) go on to argue that the government's "longer-term solution was to engage in a consultative exercise with the grains and oilseeds policy community with the objective of designing new programs to achieve income security and stability."

By contrast, the Australian government used this crisis as a catalyst to break from policies which provided direct, general assistance for farmers. In particular, the IAC and the Bureau of Agricultural Economics (BAE) argued that statutory wheat marketing created inefficiencies by distancing growers from markets and preventing them from responding to market signals (IAC 1983, 1988; BAE 1987). In addition, as the AWB collectivised wheat prices and costs, providing an average return, the IAC claimed that this approach failed to reward the most efficient growers and deprived them of incentives, instead rewarding mediocrity. This argument in particular represented a clear shift away from the collectivism of the post-Second World War years and sought to motivate policy change based on the interests of individual wheat farmers.

The rhetoric of the individual resonated with the policy ambitions of the Hawke government, expressed with particular clarity by Minister for Primary Industry John Kerin. In the 1988 policy statement "Australian Wheat Industry: Marketing in the Future", Kerin claimed that "There is a need to shape the direction of the nation's grains industry towards a future where commercial independence, maximum efficiency and marketing flexibility will be key criteria" (Commonwealth of Australia 1988, p. 2). Competition, it was claimed, would reduce growers' costs and increase the potential for higher wheat prices, which, the Department of Primary Industry (Commonwealth of Australia 1988, p. 5) argued, would benefit growers who "should be concerned with maximising their net return".

The Demise of Statutory Wheat Marketing in Australia

In 1999, the AWB was privatised (McCorriston and MacLaren 2007) despite significant opposition from wheat growers, who felt that a private organisation would be more concerned with returning benefits to shareholders rather than to wheat growers (Cockfield and Botterill 2007, p. 48). Despite these concerns, the privatised AWB Limited was floated on the Australian Stock Exchange in 2001 (McCorriston and MacLaren 2007, p. 638). The United Nations Oil-for-Food inquiry, which began in 2004, raised allegations that AWB Limited had been making payments to the Iraqi regime of Saddam Hussein for the purpose of securing wheat contracts. As a consequence, in 2005 the Australian government established a Royal Commission to investigate these claims under the oversight of Terence Cole (Cockfield and Botterill 2007, p. 44). This scandal provided the catalyst for full deregulation of the export wheat market (Cockfield and Botterill 2007).

The deregulation of the Australian export wheat market commenced with the passing of the Wheat Export Marketing Act, following the election of the Labor government in November of the previous year. Then Federal Minister for Agriculture, Forestry and Fisheries Tony Burke argued in support of deregulation, stating that "The industry needs significant reform to increase the level of competition…Only then will returns to growers be maximised" (Grattan 2008). This claim repeated arguments presented over a number of decades, portraying competition as the solution to the problems of farm viability arising from declining wheat prices and the increasing costs of inputs (Commonwealth of Australia 2008a, b; Commonwealth of Australia 1988; Kerin 1986; Mauldon and Schapper 1974). Policy makers claimed that the increased choice and freedom would enable farmers to exercise their self-interest, marketing skills, and knowledge, to obtain premium wheat prices. In this sense, wheat export market deregulation was portrayed as being primarily for the benefit of wheat farmers (Commonwealth of Australia 2008a; Productivity Commission 2000a, b, 2005; Allen Consulting 2000; Irving et al. 2000).

Throughout the 1980s and 1990s, Australian wheat farmers repeatedly opposed this policy shift (The Sun-Herald 1992; The Donald-Birchip Times 1992; National Farmer 1988; The Land 1988). For example, Harold Flett, President of the Donald District of the Victorian Farmers Federation (VFF), showed his support for the statutory marketing power of the AWB, stating that "retention of the 'single desk' seller status is critical" (The Donald-Birchip Times 1992). However, rather than address the economic argument against deregulation of the wheat industry, farm organisations criticised the subsidies provided to farmers in the European Union and United States, which they claimed undermined global markets and harmed Australian growers. Clinton Condon, speaking to The Australian (1989), stated that "The opportunity exists at the international level for governments to remove distortions which have plagued the international wheat trade…We all lose by interfering in free trade." By engaging with this discourse, ultimately, farm lobby groups (as distinct from farmers) helped legitimise the arguments around agricultural liberalisation, which were then used by policy makers to argue for wheat export market deregulation (O'Keeffe 2016b).

The Canadian Experience

The Canadian government followed a similar trajectory of wheat industry deregulation (Magnan 2015; Martin and Clapp 2015; Boaitey 2013; Coleman and Skogstad 1995). Throughout the early 1990s, critics of Canada's statutory wheat marketing arrangements mustered the political strength to make this policy a political issue (Skogstad 2005). In addition, the 1989 signing of the Canada-US Free Trade Agreement precipitated large amounts of grain being exported from Canada to the United States. This caused US grain growers to challenge the wheat board's role and existence, leading to "a virtually continuous stream of American investigations into its operations" (Skogstad 2005, p. 537). Similar to criticisms of the AWB, it was alleged that statutory marketing gave Canadian growers an unfair advantage (Skogstad 2005, p. 537).

As mentioned by Magnan (2015, p. 3), the mid-2000s saw a considerable change in policy, with the conservative government seeking to dismantle statutory wheat marketing in Canada. This shift is in addition to the termination of financial assistance, programmes, and subsidies which had supported farmers (Magnan 2015, p. 3). As in Australia, the majority of Canadian growers supported their Wheat Board. In an attempt to show support for its existence, the Canadian Wheat Board polled growers, finding that 62% supported maintenance of statutory marketing (Waldie 2011). Despite this support, the government proceeded with deregulation of the wheat export market.

Similar concerns have been identified in Australian-based research exploring growers' responses to the end of statutory wheat marketing. This research has found that farmers feel burdened with the additional workload associated with marketing (Head et al. 2011), while deregulation has undermined growers' returns, increased the risks associated with marketing, and led to farmers' experiences of disempowerment and disenfranchisement (O'Keeffe and Neave 2017; Baker 2016; O'Keeffe 2014). This is articulated by Martin and Clapp (2015, p. 553), who state that the deregulation of the Australian and Canadian wheat markets "has meant that farmers who previously counted on the marketing board to help market grain are now expected to turn to commodity exchanges and large grain traders for marketing assistance and risk management". O'Keeffe and Neave (2017) analyse how this is creating unequal power relations between traders and growers, many of whom feel disempowered and vulnerable. As discussed by O'Keeffe and Neave (2017), Martin and Clapp (2015), and Magnan (2011), the negative impacts on growers in Australia and Canada are particularly great for the small- to medium-sized producers.

Outline of This Book

In this work, I examine the deregulation of the wheat export market as an example of the broader neoliberalisation of Australian society. In Chap. 2, I outline my application of governmentality in this analysis, which I use to study the construction of truths such as efficiency, competition, the market, and the individual, which I suggest have been used to portray market liberalisation as an obvious, logical, and common-sense policy shift. I employ a poststructuralist approach to discourse analysis, which I explain in this chapter. I use this method in conjunction with a genealogical approach to research, employed to identify small discursive shifts in

Australian policy discourses from the early 1970s to the present. In doing so, I analyse how technologies of government, such as calculability, measurement, audit, cost-benefit analysis, have been used to shape what matters, by constructing knowledge and values which make the wheat industry, and its farmers, calculable, and therefore, amenable to governing.

Moreover, I argue that this shift has occurred as the role of the state has been reconstructed to facilitate the extension of private firms into aspects of Australian society previously thought of as the domain of the public. As I show in Chap. 3, policy makers have justified this shift by appealing to the belief that firms, moderated by markets, are best equipped to raise the performance of Australian industries and, therefore, improve the nation's economic performance. Connected to this shift, is the role of the citizen as consumer, as a powerful actor in the process of establishing the most successful, most efficient firms. In this role, the citizen as consumer values that which firms can provide in liberalised markets. In addition, the consumer and their welfare are prioritised over producers, such as wheat farmers.

In this context, the deregulation of the wheat market is framed as a necessary shift, by policy makers arguing that encouraging the participation of firms in this market will enhance efficiency and the wheat industry's contribution to the national economy. This policy change was made possible through a very specific construction of markets, firms, and consumers, one in which knowledge, values, and identities were reshaped to bring them into accord with the construction. The kinds of knowledge that could be counted as legitimate, I suggest, were constructed narrowly, as specifically quantitative information (such as wheat prices and costs) produced in markets. Specific policy values, such as efficiency, competition, and individualism, were privileged over potentially competing values, such as equity, stability, and security. Particular constructions of the identities of firms, consumers, and farmers are also integral to this policy change. For example, large, transnational firms were portrayed as integral to maximising the performance of the wheat industry, through their superior efficiencies, scale, and responsiveness to consumer needs. In this reality, wheat export market deregulation was portrayed by policy makers as common sense.

However, as I argue in Chaps. 4 and 5, this construction needed to be made operational. This was achieved through assemblages of technologies, such as audit, cost-benefit analysis, performance objectives, and econometric modelling, ostensibly introduced to make the AWB more accountable. However, these technologies of government were used to govern the industry by establishing the maximisation of wheat growers' returns as the only legitimate purpose of statutory wheat marketing policy. In addition, technologies of agency were employed through policy instruments to construct "good" farming behaviour in terms of actions associated with entrepreneurial, self-reliant individuals, who were primarily concerned with maximising their returns. Thus, consistent with this construction, policy makers justified wheat export market deregulation as being in farmers' interests.

As I argue in Chaps. 6 and 7, this portrayal did not adequately consider firms' capacity to exert power in markets. Underpinned by policy makers' conception of competition in terms of contestability, which is concerned with potential, rather than actual, competition, wheat export market deregulation has resulted in consolidated markets featuring transnational firms such as Cargill, ADM, and Glencore. These firms have sought to develop their position in the Australian wheat export market through influencing policy making and engaging in supply chain management strategies such as mergers and acquisitions, to build and protect market share.

Chapter 8 draws this research together and examines the key themes emerging in this work. I draw parallels between the deregulation of the wheat export market, and other areas of Australian policy, to critically reflect on the rationality which has underpinned the neoliberalisation of Australian society. I challenge the notion of the "business-friendly environment" as a necessary precursor to a prosperous and happy society, and instead argue that this construction has been used to transfer power from the public to the private sphere.

References

Allen Consulting, (2000), *The Wheat Marketing Act 1989 – The Economic Impact of Competitive Restrictions*, Sydney: Allen Consulting Group.

Alston, J., (1986), Consequences of deregulation in the Victorian egg industry, *Review of marketing and agricultural economics*, 54(1), pp. 33–43.

Alston, M., (2012), Rural Male Suicide in Australia. *Social Science and Medicine*, 74, pp. 515–522.

Alston, M., (2004), Who is Down on the Farm? Social aspects of Australian agriculture in the 21st century, *Agriculture and Human Values*, 21, pp. 37–46.

Anderson, K., Lloyd, P., and MacLaren, D., (2007), Distortions to agricultural incentives in Australia since World War II, *Economic Record*, 83(263), pp. 461–482.

Argent, N., (2011), Trouble in paradise? Governing Australia's multifunctional rural landscapes, *Australian Geographer*, 42(2), pp. 183–205.
Argent, N., (2005), The neoliberal seduction: Governing-at-a-distance, community development and the battle over financial services provision in Australia, *Geographical Research*, 43(1), pp. 29–39.
Argent, N., (2002), From pillar to post? In search of the Post-Productivist Countryside in Australia, *Australian Geographer*, 33(1), pp. 97–114.
Australian Bureau of Agricultural and Resources Economics and Sciences, (2012), *Agricultural Commodity Statistics 2012: Table 201 – Australian wheat area, yield and production, by state*, [online] http://www.daff.gov.au/abares/publications [Accessed November 11 2015].
Australian Bureau of Statistics, (2012), *Australian farming and farmers, Australian Social Trends*, December 2012, No. 4102.0, [online] http://www.abs.gov.au/AUSSTATS/abs@.nsf/Lookup/4102.0Main+Features10Dec+2012 [Accessed November 6 2017].
Baker, C., (2018), The nation building state retreats: An Australian case study in the changing role of the state, *Journal of Rural Studies*, 62, pp. 146–155.
Baker, C., (2016), *Economic rationalism and Australian agriculture: the nation-building state retreats*, Paper presented at the workshop: From economic rationalism to global neoliberalism, December 2nd 2016, RMIT University.
Beer, A., Clower, T., Haughton, G., and Maude, A., (2005), Neoliberalism and the institutions for regional development in Australia, *Geographical Research*, 43(1), pp. 49–58.
Bjørkhaug, H., and Richards, C., (2008), Multifunctional agriculture in policy and practice? A comparative analysis of Norway and Australia, *Journal of Rural Studies*, 24, pp. 98–111.
Boaitey, A., (2013), Grain market deregulation: a case study of the Canadian and Australian wheat boards, *Journal of Public Affairs*, 13(3), pp. 282–287.
Botterill, L., (2012), *Wheat marketing in transition*, New York: Springer.
Bradshaw, B., (2004), Plus c'est la meme chose? Questioning crop diversification as a response to agricultural deregulation in Saskatchewan, Canada, *Journal of Rural Studies*, 20, pp. 35–48.
Brasier, K., (2002), Ideology and discourse: Characterisations of the 1996 Farm Bill by agricultural interest groups, *Agriculture and Human Values*, 19, pp. 539–253.
Breustedt, G., and Glauben, T., (2007), Driving forces behind exiting from farming in Western Europe, *Journal of Agricultural Economics*, 58(1), pp. 115–127.
Bryant, L., & Garnham, B., (2014), Economics, ethics and emotions: Farmer distress within the moral economy of agribusiness. *Journal of Rural Studies* 34, pp. 304–312.
Bureau of Agricultural Economics, (1987), *Wheat marketing and assistance*, Canberra: Australian Government Publishing Service.

Caldwell, T., Jorm, A., and Dear, K., (2004), Suicide and Mental Health in Rural, Remote and Metropolitan Areas in Australia, *MJA*, 181(7), pp. 10–14.

Cheshire, L., Meurk, C., and Woods, M., (2013), Decoupling farm, farming and place: Recombinant attachments of globally engaged family farmers, *Journal of Rural Studies*, 30, pp. 64–74.

Cheshire, L., and Lawrence, G., (2005), Neoliberalism, Individualisation and Community: Regional Restructuring in Australia, *Social Identities*, 11(5), pp. 435–445.

Cockfield, G., and Botterill, L., (2007), Deregulating Australia's wheat trade: From the Australian Wheat Board to AWB Limited, *Public Policy*, 2(1), pp. 44–57.

Cockfield, G., and Botterill, L., (2006), Rural adjustment Schemes: Juggling politics, welfare and markets, *Australian Journal of Public Administration*, 65(2), pp. 70–82.

Cocklin, C., and Dibden, J., (2002), Taking stock: Farmers' reflections on the deregulation of Australian dairying. *Australian Geographer*, 33(1), 29–42.

Cocklin, C., and Dibden, J., (2005), *Sustainability and change in rural Australia*, Sydney: University of New South Wales Press.

Coleman, W., and Skogstad, G., (1995), Neo-liberalism, policy networks and policy change: Agricultural policy reform in Australia and Canada, *Australian Journal of Political Science*, 30(2), pp. 242–263.

Commonwealth of Australia, (2012), *Parliamentary debates.* House of Representatives: official Hansard, Wednesday 31 October.

Commonwealth of Australia, (2008a), *Parliamentary debates.* House of Representatives: official Hansard, Thursday 13 March.

Commonwealth of Australia, (2008b), *Parliamentary debates.* Senate: official Hansard, Monday 16 June.

Commonwealth of Australia, (1988), *Australian wheat industry: Marketing in the future*, Canberra: Department of Primary Industries and Energy.

Davidson, A., (2002), Globalisation, uneven development and marginalisation: Dairy restructuring in New South Wales, *Australian Journal of Social Issues*, 37(1), pp. 85–102.

Davidson, A., (2001), Dairy deregulation and the demise of dairying in New South Wales, *Rural Society*, 11(1), pp. 23–37.

Dean, M., (1999), *Governmentality: Power and Rule in Modern Society*, London: Sage Publications.

Dean, M., (1998), Risk, calculable and incalculable, *Soziale Welt*, 49, pp. 25–42.

Dean, J., and Stain, H., (2010), Mental health impact for adolescents living with prolonged drought, *Australian Journal of Rural Health*, 18, pp. 32–37.

Dibden, J., Potter, C., and Cocklin, C., (2009), Contesting the neoliberal project for agriculture: Productivist and multifunctional trajectories in the European Union and Australia, *Journal of Rural Studies*, 25(3), pp. 299–308.

Dixon, D., and Hapke, H., (2003), Cultivating discourse: The social construction of agricultural legislation, *Annals of the Association of American Geographers*, 93(1), pp. 142–164.

The Donald-Birchip Times, (1992), *Grains Comment*, July 10 1992, Public Record Office of Victoria, VA 1057.

Dufty-Jones, R., (2015), Governmentalities of mobility: The role of housing in the governance of Australian rural mobilities, *Journal of Rural Studies*, 42, pp. 63–78.

Edwards, G., (1987), Agricultural policy debate: a survey, *The Economic Record*, June, pp. 129–143.

Fairbairn, M., (2014), 'Like gold with yield': Evolving intersections between farmland and finance, *Journal of Peasant Studies*, 41(5), pp. 777–795.

Fraser, C., Smith, K., Judd, F., Humphreys, J., Fragar, L., and Henderson, A., (2005), Farming and Mental Health Problems and Mental Illness, *International Journal of Social Psychiatry*, 51, pp. 340–349.

Gibson, C., Dufty, R., Phillips, S., and Smith, H., (2008), Counter-geographies: The campaign against rationalisation of agricultural research stations in New South Wales, Australia, *Journal of Rural Studies*, 24, pp. 351–366.

Gosling, E., and Williams, K., (2010), Connectedness to nature, place attachment and conservation behaviour: Testing connectedness theory among farmers, *Journal of Environmental Psychology*, 20, pp. 298–304.

Grattan, M., (2008), *Split over wheat desk reform*, The Age, March 14 2008, [online] http://www.theage.com.au/news/business/split-over-wheat-desk-reform/2008/03/13/1205126110679.html [Accessed October 26 2016].

Gray, I., and Lawrence, G., (2001a), *A future for regional Australia: Escaping global misfortune*, Cambridge: Cambridge University Press.

Gray, I., and Lawrence, G., (2001b), Neoliberalism, individualism and prospects for regional renewal, *Rural Society*, 11(3), pp. 283–298.

Gray, I., and Lawrence, G., (1996), Predictors of stress among Australian farmers, *Australian Journal of Social Issues*, 31(2), pp. 173–189.

Hawke, B., (1986), *A message from the Prime Minister*, The Mail-Times, February 28 1986.

Head, L., Atchison, J., and Gates, A., (2012), *Ingrained: A human bio-geography of wheat*, Surrey: Ashgate Publishing Limited.

Head, L., Atchison, J., Gates, A., and Muir, P., (2011), A fine-grained study of the experience of drought, risk and climate change among Australian wheat farming households, *Annals of the Association of American Geographers*, 101(5), pp. 1089–1108.

Herbert-Cheshire, L., (2000), Contemporary strategies for rural community development in Australia: a governmentality perspective, *Rural Society*, 16, pp. 201–215.

Higgins, V., (2005), *Governing agriculture through the managerial capacities of farmers*, in Higgins, V., and Lawrence, G., (eds), Agricultural Governance: Globalisation and the new politics of regulation, pp. 118–132, London: Routledge.

Higgins, V., (2002a), *Constructing reform: Economic expertise and the governing of agricultural change in Australia*, New York: Nova Science Publishers.

Higgins, V., (2002b), Self-reliant citizens and targeted populations: The case of Australian agriculture in the 1990s, *Arena Journal*, 19, pp. 161–177.

Higgins, V., (2001a), Assembling restructuring: governmentality, economic regulation and the historical emergence of the 'enterprising farmer' in Australian agricultural policy, *Review of International Political Economy*, 8(2), pp. 311–328.

Higgins, V., (2001b), Governing the boundaries of viability: Economic expertise and the production of the 'low income farm problem' in Australia, *Sociologia Ruralis*, 41(3), pp. 358–375.

Higgins, V., and Lockie, S., (2002), Re-discovering the social: neo-liberalism and hybrid practices of governing in rural natural resource management, *Journal of Rural Studies*, 18, pp. 419–428.

Industry Assistance Commission, (1988), *The Wheat Industry*, Report No. 411, AGPS, Canberra.

Industry Assistance Commission, (1983), *The Wheat Industry*, Report No. 329, AGPS, Canberra.

Industry Commission, (1991), *Statutory marketing arrangements for primary products*, Report No. 10, Canberra: Australian Government Publishing Service.

Industries Assistance Act 1973 (Commonwealth)

Irving, M., Arney, J., and Linder, B., (2000), *National Competition Policy review of the Wheat Marketing Act 1989*, Canberra: National Competition Policy – Wheat Marketing Act Review Committee.

Kerin, R., (1986), *Poor prospects ahead*, The Mail-Times, February 28 1986.

Kleijn, D., and Sutherland, W., (2003), How effective are European agri-environment schemes in conserving and promoting biodiversity?, *Journal of Applied Ecology*, 40, pp. 947–969.

Kuehne, G., (2012), My Decision to Sell the Family Farm, *Agriculture and Human Values* 30(2), pp. 1–11.

The Land, (1988), *Growers say leave AWB as sole exporter*, June 30 1988.

Larner, W., (2000), Post-welfarist state governance: Towards a code of social and family responsibility, *Social Politics* 7(2): 244–265.

Lawrence, G., (1987), *Capitalism and the countryside: The rural crisis in Australia*, Sydney: Pluto Press.

Lawrence, G., Richards, C., and Lyons, K., (2013), Food security in Australia in an era of neoliberalism, productivism and climate change, *Journal of Rural Studies*, 29(1), pp. 30–39.

Lazonick, W., (2012), The financialisation of the US corporation: What has been lost, and how it can be regained, *Seattle University Law Review*, 36, pp. 857–909.

Le Heron, R., and Roche, M., (1999), Rapid regulation, agricultural restructuring, and the reimaging of agriculture in New Zealand, *Rural Sociology*, 64(2), pp. 205–218.

Liepins, R., (2000), Exploring rurality through 'community': Discourses, practices and spaces shaping Australian and New Zealand rural 'communities', *Journal of Rural Studies*, 16, pp. 325–341.

Liepins, R., (1996), Reading agricultural power: Media as sites and processes in the construction of meaning, *New Zealand Geographer*, 52(2), pp. 3–10.

Liepins R, and Bradshaw, B., (1999), Neo-liberal agricultural discourse in New Zealand: Economy, culture and politics linked, *Sociologia Ruralis* 39(4), pp. 563–582.

Lobao, L., and Meyer, K., (2001), The great agricultural transition: Crisis, change and social consequences of Twentieth Century US farming, *Annual Review of Sociology*, 27, pp. 103–124.

Lobao, L., and Stofferahn, C., (2008), The community effects of industrialised farming: Social science research and challenges to corporate farming laws, *Agriculture and Human Values*, 25, pp. 219–240.

Lockie, S., (1999), The state, rural environments, and globalisation: 'action at a distance' via the Australian Landcare program, Environment and Planning A, 31, pp. 597–611.

Lockie, S., and Bourke, L., (eds.) (2001), Rurality Bites, Pluto Press: Annandale, New South Wales.

Lockie S, and Higgins, V., (2007), Roll-out neoliberalism and hybrid practices of regulation in Australian agri-environmental governance, *Journal of Rural Studies*, 23(1), pp. 1–11.

Magnan, A., (2015), The financialisation of agri-food in Canada and Australia: Corporate farmland and farm ownership in the grains and oilseed sector, *Journal of Rural Studies*, 41, pp. 1–12.

Magnan, A., (2011), *The limits of farmer-control: Food sovereignty and conflicts over the Canadian wheat board*, in Desmarais, A., Wiebe, N., and Wittman, H., (eds), Food Sovereignty in Canada: Creating just and sustainable food systems, pp. 114–133, Black Point: Fernwood Publishing.

Margetts, D., (2007), National Competition Policy and the Australian dairy industry, *Journal of Australian Political Economy*, 60, pp. 98–129.

Martin, S., and Clapp, J., (2015), Finance for agriculture or agriculture for finance?, *Journal of Agrarian Change*, 15(4), pp. 549–559.

Mather, C., and Greenberg, S., (2003), Market liberalisation in post-Apartheid South Africa: The restructuring of citrus exports after 'deregulation', *Journal of Southern African Studies*, 29(2), pp. 393–412.

Mauldon, R., and Schapper, H., (1974), *Australian farmers under stress in prosperity and recession*, Perth: University of Western Australia Press.

McCorriston, S., and MacLaren, D., (2007), Deregulation as (welfare reducing) trade reform: The case of the Australian Wheat Board, *American Journal of Agricultural Economics* 89(3) pp. 637–650.

McKenzie, F., (1994), Population Decline in Non-Metropolitan Australia: Impacts and Policy Implications, *Urban Policy and Research*, 12(4), pp. 253–264.

McManus, P., Walmsley, J., Argent, N., Baum, S., Bourke, L., Martin, J., Pritchard, B., and Sorenson, T., (2012), Rural community and rural resilience: What is important to farmers in keeping their country towns alive?, *Journal of Rural Studies*, 28, pp. 20–29.

Miller, P., and Rose, N., (2008), *Governing the present: Administering economic, social and personal life*, Cambridge: Polity.

Miller, P., and Rose, N., (1990), Governing economic life, *Economy and Society*, 19(1), pp. 1–31.

Morgan, G., (2014), Financialisation and the modern corporation, *Transfer*, 20(2), pp. 183–197.

Mugera, A., Curwen, R., and White, B., (2016), Deregulation of the Australian wheat export market: What happened to wheat prices?, *Journal of International Food and Business Marketing*, 28(1), pp. 18–34.

Musgrave, W., (1979), Rural poverty: Policy options, *Review of Marketing and Agricultural Economics*, 47(3), pp. 191–205.

Musgrave, W., (1990), *Rural Adjustment*, in Williams, D., (ed.), Agriculture in the Australian Economy, Sydney: Sydney University Press.

National Farmer, (1988), *National Farmer survey*, July 1 1988, Public Records Office Victoria, Melbourne.

O'Keeffe, P., and Neave, M., (2017), Experiences of wheat growers in Victoria's western Wimmera following deregulation of the Australian wheat export market, *Rural Society*, 26(1), pp. 1–18.

O'Keeffe, P., (2016a), *Responding to the productivist paradigm: Experiences of farmers in Victoria's western Wimmera*, in M. Chou (ed.) Proceedings of the Australian Sociological Association (TASA) 2016 Conference, Melbourne, Australia, 28 November – 1 December 2016, pp. 263–269.

O'Keeffe, P., (2016b), *Towards deregulation: Shifting discourses in the Australian wheat industry*, in 2016 Asia Pacific Economic and Business History Conference Papers, Adelaide, Australia, 11–13 February 2016, pp. 1–25.

O'Keeffe, P., (2014), *The impact of wheat export market deregulation upon growers in Victoria's western Wimmera*, in Proceedings of the 58th National Australian Agricultural and Resource Economics Society Annual Conference (AARES 2014), Canberra, Australia, 4–7 February 2014, pp. 1–38.

Peck, J., and Tickell, A., (2002), Neoliberalising space, *Antipode*, June 2002, pp. 380–404.

Pietola, K., Väre, M., and Lansink, A., (2003), Timing and type of exit from farming: farmers' early retirement programmes in Finland, *European Review of Agricultural Economics*, 30(1), pp. 99–116.

Potter, C., and Tilzey, M., (2007), Agricultural multifunctionality, environmental sustainability and the WTO: Resistance or accommodation to the neoliberal project for agriculture?, *Geoforum*, 38, pp. 1290–1303.

Potter, C., and Tilzey, M., (2005), Agricultural policy discourses in the European post-Fordist transition: neoliberalism, neomercantilism and multifunctionality, *Progress in Human Geography*, 29(5), pp. 581–600.

Pretty, G., Chipuer, H., and Bramston, P., (2003), 'Sense of place among adolescents and adults in rural Australian towns: The discriminating features of place attachment, sense of community and place dependence in relation to place identity', *Journal of Environmental Psychology*, 23, pp. 273–287.

Pritchard, B., Argent, N., Baum, S., Bourke, L., Martin, J., McManus, P., Sorenson, A., and Walmsley, J., (2012), Local – If possible: How the spatial networking of economic relations amongst farm enterprises aids small town survival in rural Australia, *Regional Studies*, 46(4), pp. 539–557.

Pritchard, B., Burch, D., & Lawrence, G., (2007), Neither 'family' nor 'corporate' farming: Australian tomato growers as farm family entrepreneurs, *Journal of Rural Studies*, 23, pp. 75–87.

Pritchard, B., (2006), The political construction of free trade visions: The geopolitics and geo-economics of Australian beef exporting, *Agriculture and Human Values*, 23, pp. 37–50.

Pritchard, B., (1999), National Competition Policy in action: The politics of agricultural deregulation and wine grape marketing in the Murrumbidgee Irrigation Area, *Rural Society*, 9(2), pp. 421–442.

Pritchard, W., (1998), The emerging contours of the Third Food Regimes: Evidence from Australian Dairy and Wheat Sectors, *Economic Geography*, 74(1), pp. 64–74.

Productivity Commission, (2017), *Regulation of Australian agriculture*, Productivity Commission Inquiry Report, No. 79, [online] https://www.pc.gov.au/inquiries/completed/agriculture/report/agriculture.pdf [Accessed November 6 2017].

Productivity Commission, (2005), *Review of National Competition Policy reforms: Productivity Commission Inquiry Report*, No. 33, 28 February 2005, Commonwealth of Australia, Canberra.

Productivity Commission, (2000a), *Single-desk marketing: Assessing the economic arguments*, [online] http://www.pc.gov.au/research/supporting/single-desk [Accessed December 20 2016].

Productivity Commission, (2000b), *Productivity Commission submission to National Competition Policy review of the Wheat Marketing Act 1989*, Melbourne: Productivity Commission.

Raymond, C., Bieling, C., Fagerholm, N., Martin-Lopez, B., and Plieninger, T., (2016), The farmer as a landscape steward: Comparing local understandings of landscape stewardship, landscape values, and land management actions, *Ambio*, 45, pp. 173–184.

Rose, N., (1996), The death of the social? Re-figuring the territory of government, *Economy and Society*, 25(3): 327–356.

Rose, N., (1993), Government, authority and expertise in advanced liberalism, *Economy and Society*, 22(3), pp. 283–299.

Royal Commission into Grain Storage, Handling and Transport, (1988), *Royal Commission into Grain Storage, Handling and Transport*, Canberra: AGPS.

Ryan, T., (1984), Wheat Marketing, *Review of Marketing and Agricultural Economics*, 52(2), pp. 117–128.

Sharp, L., and Richardson, T., (2001), Reflections on Foucauldian discourse analysis in planning and environmental policy research, *Journal of Environmental Policy and Planning*, 3, pp. 193–209.

Skogstad, G., (2005), The dynamics of institutional transformation, The case of the Canadian Wheat Board, *Canadian Journal of Political Science*, September, pp. 529–548.

Skogstad, G., (1998), Ideas, paradigms and institutions: Agricultural exceptionalism in the European Union and the United States, *Governance: An international Journal of Policy and Administration*, 11(4), pp. 463–490.

Smailes, P., (2000), The diverging geographies of social and business interaction patterns: a case study of rural South Australia, *Geographical Research*, 38 (2), pp. 158–181.

Smith, E., and Pritchard, B., (2012), *Australian Agricultural and Rural Policy since World War II: The pursuit of agricultural efficiency*, Scoping a Vision for the Future for Rural and Regional Australia: Collection of Papers presented at the Sustaining Rural Communities Conference 2012, ANU College of Arts and Social Science.

Stilwell, F., (1992), *Understanding cities and regions: Spatial political economy*, Leichhardt: Pluto Press.

The Sun-Herald, (1992), *Country view: From the NSW Farmers Federation*, July 12 1992.

Talbot, L., and Walker, R., (2007), Community perspectives on the impact of policy change on linking social capital in a rural community, *Health and Place*, 13, pp. 482–492.

Tennent, R., and Lockie, S., (2013), Vale Landcare: The rise and decline of community-based natural resource management is rural Australia, *Journal of Environmental Planning and Management*, 56(4), pp. 572–587.

Tonts, M., (1999), Some recent trends in Australian regional economic development policy, *Regional Studies*, 33(6), pp. 581–586.

Tonts, M., and Haslam-McKenzie, F., (2005), Neoliberalism and changing regional policy in Australia, *International Planning Studies*, 10(3), pp. 183–200.

Tonts, M., and Jones, R., (1997), From State Paternalism to Neoliberalism in Australian Rural Policy: Perspectives from the Western Australian wheatbelt, *Space and Polity*, 1(2), pp. 171–190.

Vanclay, F., (2003), The Impacts of Deregulation and Agricultural Restructuring For Rural Australia, *Australian Journal of Social Issues*, 38(1), pp. 81–92.

Vanclay, F., and Lawrence, G., (1993), Environmental and Social Consequences of Economic Restructuring in Australian Agriculture, *International Journal of Sociology of Agriculture and Food*, 3, pp. 97–118.

Waldie, P., (2011), *Prairie farmers vote to keep Canadian Wheat Board*, The Global and Mail, September 12 2011, [online] https://beta.theglobeandmail.com/report-onbusiness/ prairie-farmers-vote-to-keep-canadian-wheatboard/article 556370/?ref=http://www.theglobeandmail.com& [Accessed November 6 2017].

Walker, A., and Ryan, R., (2008), Place attachment and landscape preservation in rural New England: A Maine case study, *Landscape and Urban Planning*, 86, pp. 141–152.

Warhurst, J., (1982), The Industries Assistance Commission and the making of primary industry policy, *Australian Journal of Public Administration*, 16(1), pp. 15–32.

Weiss, C., (1999), Farm growth and survival: Econometric evidence for individual farms in upper Austria, *American Journal of Agricultural Economics*, 81, pp. 103–116.

Williams, J., (2012), *Grain price volatility – a problem, a necessity, and cause for reform*, Paper selected for presentation and the IFAMA 22nd Annual World Symposium, 'The Road to 2050: The China Factor', Shanghai, China, June 11–12 2002.

Wiseman, L., and Whiteford, G., (2009) Understanding occupational transitions: A study of older rural men's retirement experiences, *Journal of Occupational Science* 16(2), pp. 104–109.

Wittmaack, N., (2006), Should corporate farming be limited in the United States?: An economic perspective, *Major Themes in Economics*, Spring, pp. 45–59.

Woods, M., (2006), Redefining the 'rural question': The new 'politics of the rural' and social policy, *Social Policy and Administration*, 40(6), pp. 579–595.

Wythes, A., and Lyons, M., (2006), Leaving the land: An exploratory study of retirement for a small group of Australian men, *Rural and Remote Health*, 6, pp. 531–540.

CHAPTER 2

Governmentality as a Lens for Analysing Agricultural Restructuring in Australia

Introduction

The case of Australian wheat marketing policy, across the past 30 years, is a particularly interesting one. The industry has ostensibly been deregulated, implying the withdrawal of the state. It could be argued that the deregulation of the wheat export market is made possible through the simple retraction of state support. However, deregulation is not so easily explained. Rather than conceptualising deregulation in terms of a simple absence of regulation, I draw on governmentality research and its associated analysis of how governmental technologies of agency and performance have been employed to make society amenable to governing. I explore the construction of individuals—in particular, farmers—as well as policy values and organisations as part of this changing environment.

Australian policy makers' attempts to increase efficiency and productivity through quantifiable indicators such as prices and costs are not confined to the Australian wheat industry. Similar processes have been applied to restructure labour markets and diverse sectors such as health care, social security, and higher education, in Australia and globally (Watts 2016; Connell 2013; Cooper and Allem 2008, Alston 2007). Central to these policy shifts is the construction of markets as "object[s] of power and knowledge" (Mitchell 2008, p. 1117). According to this construction, evident in mainstream Australian policy discourses, markets and economies exist naturally and in reality as apolitical mechanisms for the efficient ordering of the economy through transactions, "in

P. O'Keeffe, *Making Markets in Australian Agriculture*,
https://doi.org/10.1007/978-981-13-3519-8_2

the name of national and individual prosperity" (Rose 1993, p. 286). This rationality has been applied by policy makers in the economic restructuring of Australian society (Watts 2016; Alston 2004, 2007; Pritchard 2005). However, the rationality of liberalised markets and economies alone is not sufficient to make this shift possible. Society must be made governable, and amenable to this reality of markets and the economy (Gibson, Dufty, Phillips, and Smith 2008; Oels 2005; O'Malley, Weir, and Shearing 1997).

Sociology of quantification offers a useful approach here, as a lens for analysing how Australian farmers and farming have been made calculable and administrable. I draw on sociology of quantification research, in conjunction with governmentality literature on the technology of calculability, to analyse the construction and prioritisation of quantitative information as legitimate knowledge. This form of knowledge has been prioritised in policy making, while at the same time, the social world has effectively been erased, as I argue in Chap. 5. Quantification produces knowledge which can be simplified and easily communicated, and therefore able to be acted upon. In addition, this is knowledge which in many cases is produced in markets. Thus, quantification of social phenomena makes society, including industries such as farming and people such as farmers, amenable to governing. However, technologies of government, such as audit, performance measures, benchmarking, and agency, help operationalise the rationalities of markets, firms, and individuals. Such technologies help shape individuals and organisations, creating entrepreneurial, self-reliant actors who value choice and the freedoms offered by markets, and who understand and respond to the quantified knowledge produced in markets.

In this chapter, I outline my use of governmentality throughout this book, specifically to understand the shaping of knowledge and farmers' identities, as a means of constructing a governable reality. I then explain the relationship between this approach and my use of discourse analysis in understanding truth-making and the erasure of competing discourses in Australian agricultural policy. Following this, I draw on research in the sociology of quantification field, before turning to literature analysing technologies of agency and performance, to help build a theoretical framework for this research. Finally, I outline the genealogical approach used in this research, to explore the critical shifts which have helped make wheat export market deregulation possible.

Governmentality: An Introduction

Governmentality literature draws from Foucault's concept of government as "the conduct of conduct" (Argent 2005, p. 30; Dean 1998, p. 26). In analysing neoliberal governance, this body of work seeks to understand the relationship between the rationalities of the market, and the individual, for example, and the technologies which render these rationalities thinkable, knowable, and governable (Lockie and Higgins 2007; Dean 1998). This research seeks to understand the practices which shape of individual actions, and that of the population, towards particular political aims, economic goals, and ideologies (Dean 1998). As stated by Argent (2005, p. 30), this reflects changed approaches to governing, as governments have become increasingly concerned with the question of "how to best inculcate within a national population the appropriate moral code, mode of behaviour and standards of comportment compatible with the common weal of emerging capitalist, essentially monotheist, societies". In this sense, advanced liberal governing seeks to governing through society, by encouraging behaviours, values, and practices which best comply with capitalist societies (Rose 1996). Understanding how knowledge, behaviours, and values of individuals and communities are shaped is critically important. However, governing through society also requires people to view themselves as individuals and who accept risk and responsibility as being individualised. In additional, this requires individuals to act in ways which intend to minimise their exposure to risk. That is, to act rationally, economically, and in an entrepreneurial manner (Herbert-Cheshire and Higgins 2004).

Governmentality research analyses the rationalities, and the seemingly mundane technologies of agency and performance, which make this approach to governing possible (Higgins 2001a; Rose 1999; Miller and Rose 1990). According to Rose (1999, p. 22), an analytics of government seeks to:

> […] interrogate the problems and problematizations through which 'being' has been shaped in a thinkable and manageable form, the sites and locales where these problems formed and the authorities responsible for enunciating upon them, the techniques and devices invented, the modes of authority and subjectification engendered, and the telos of these ambitions and strategies.

As mentioned by Rose (1993, p. 286), whereas the object of the economy, featuring markets comprised of competing firms, may not be dissimilar to other modes of governing such as liberalism, advanced liberal governing employs technical, practical technologies to make society amenable to that rationality. According to Miller and Rose (1990, p. 3), governmentality research entails:

> [...] an investigation not merely of grand political schema, or economic ambitions, nor even of general slogans such as 'state control', nationalization, the free market and the like, but of apparently humble and mundane mechanisms which appear to make it possible to govern: techniques of notation, computation and calculation.

In a sense, attention therefore shifts from "the state", to the "plurality of governing bodies", and the rationalities, technologies, and practices which are used to govern, to shape the behaviours, values, and practices of individualities and populations (Dean 1998, p. 26). As mentioned by Higgins (2001a), these technologies may involve methods of accounting, Acts of Parliament, and government White Papers, though also could include technologies such as auditing, benchmarking, the use of performance objectives, and econometric modelling. These technologies share the ambitions of making things knowable, through reductive and calculative technologies, for example, and then amenable to governing, through technologies which seek to measure performance of communities and individuals, according to that knowledge. This knowledge might be claimed to be the result of neutral technologies. However, as Argent (2005, p. 31) claims, governmentality studies "seek to situate advanced liberal (or neoliberal) modes of government historically by revealing the techniques by which their agencies of authority create and disseminate normative 'truths' in accordance with their particular moral vision".

Identity and the Individual

In particular, governmentality research has focused on the creation and governance of self-governing individuals and the subtle coercion of these individuals through technologies of agency and performance (O'Keeffe 2017a, b; Penny 2016; Higgins et al. 2015; Lockie 2009; Higgins and Lockie 2002). This work analyses the construction of people as individuals, who are economically rational, self-interested, and self-reliant, and who seek to maximise their utility through choices in markets. Using a governmentality lens, Higgins (2002a) and Lockie and Higgins (2007,

p. 3) claim that citizenship has been framed in terms of an individual's capacity to be entrepreneurial. To be a social responsible citizen, in this construction, is to engage in good economic decision making, which carefully considers risks and costs. This requires individuals who communicate and interpret market signals, and base decisions on calculation. In turn, the entrepreneurial citizen understands risk and responsibility concerning their decisions, and accepts these risks and responsibilities as their own (Emery 2015; Cheshire and Lawrence 2005; Herbert-Cheshire and Higgins 2004). Furthermore, the entrepreneurial citizen is the one who demands freedom and has the capacity to make choices in markets (Higgins 2002a, b). This freedom is often connected to individual empowerment, however, as Higgins (2002b, p. 165) argues:

> Here freedom is not a domain of liberty that is separate from power but rather the effect of a particular form of power. Experts are significant in 'empowering' individuals to rationally manage their life though certain professionally ratified mental, ethical and practical techniques for active self-management.

In this regard, this construction of freedom is a regulated freedom, which is restricted to practices and behaviours which are deemed to be legitimate, according to the boundaries established by policy discourses.

Thus, the entrepreneurial, rational, and self-reliant actor is critical in enabling governing through society to be made possible. This construction occurs in the problematisations of the state as inefficient, restrictive, and over-bearing. Power, it is claimed, is shifted to the individual, through markets which enabled the entrepreneurial individual to act rationally, make choices, and exercise their freedom, in the assumption that this will be exercised in their own self-interest. However, as Rose (1993, 1996), Larner (2000), Higgins (2002b), and Argent (2005) suggest, this is a critical element in making governing in an advanced liberal way possible. The notion of the individual and its primacy then enables the individual to be governed surreptitiously, through governmental technologies that shape legitimate ways of knowing, acting, and thinking. In other words, this form of governing is enacted "through the regulated and accountable choices of autonomous agents" (Rose 1993, p. 298). Thus, the individual and their identity, values, and actions are critically important to advanced liberal governing.

Making People Knowable

The calculating, entrepreneurial individual is significant in this mode of governing, as is the construction of individuals and communities who are calculable. Here, I turn to the extension of individualism and self-reliance to rural community development projects, before exploring the ways in which people and communities are made calculable, and how this is made possible through technologies of government.

Australian agricultural policy in recent decades has focused on promoting individualism, competition, self-help, and self-reliance (Herbert-Cheshire and Lawrence 2005; Cheshire and Lawrence 2005; Gray and Lawrence 2001; Higgins 2001a). I extend this analysis by exploring the dominant conceptualisation of farmers as business-minded individuals who are centred exclusively in maximising their returns. This construction differs substantially from research analysing the importance of farmers' connection to their land, their family histories in farming, and the relationship between farming, identity, and well-being (McManus et al. 2012; Kuehne 2012; Sartore et al. 2008; Fraser et al. 2005; Smailes 2000; Gray and Lawrence 1996). Policy discourses have constructed farming in economic terms, largely externalising these important social aspects of farming, rendering them insignificant in policy. As I will argue, this discursive shift has helped to make deregulation possible.

However, as Herbert-Cheshire and Higgins (2004, p. 290) highlight, this shift is also dependent upon the use of technologies which help make individuals and communities knowable. As Herbert-Cheshire and Higgins (2004, p. 290) argue, "expertise seeking to achieve change in an advanced liberal way acts as a key 'centre of calculation' in making 'community' knowable, and in constituting the discursive framework through which communities can reflect on their conduct and transform themselves into active agents in their self-governance." In this regard, governing in an advanced liberal way requires people who not only understand themselves as individuals who make decisions based on calculation but also understand and regulate their own conduct accordingly. That is, people who rationalise their own actions, calculate, reflect, and then act according to calculation. This requires people to see themselves, their value, and knowledge concerning their performance in terms of how it is calculated. In this work, I explore the construction and governance of individual farmers in Australia. Before governing through individuals is possible, I argue those individuals must be rationalised and made calculable. This, I suggest, requires the discursive construction of a reality that portrays this rationalisation as being common sense, and inherently good policy. I now turn to poststructuralist approaches to studying discourse, as a method of analysing truth-making and the construction of a reality which prioritises marketisation, efficiency, and productivity, on the one hand, and facilitates the decoupling of farmers from the land and communities, on the other.

Discourse and the Shaping of "Truth" and Reality

To understand how liberalisation of the wheat export market came to be presented in policy discourses as a common-sense, logical policy shift, I analyse discursive constructions which produce and reproduce truths such as competition, efficiency, productivity, freedom, choice, and individualism. Similar approaches have been applied to policy studies in the United States (Dixon and Hapke 2003; Brasier 2002), New Zealand (Liepins and Bradshaw 1999; Liepins 1996), the United Kingdom (Potter and Tilzey 2005; Stenson and Watt 1999), and Australia (Lockie and Higgins 2007; Pritchard 2005a, b; Tonts and Haslam-McKenzie 2005; Herbert-Cheshire 2000). These studies explore policy discourses to understand how meaning and knowledge is created, and how this is exercised as a form of power through discourse. In particular, Dibden, Potter, and Cocklin (2009), Higgins (2001a, b, 2002a, b, 2005), and Lockie and Higgins (2007) employ a poststructuralist approach to discourse analysis, to understand particular policies and policy shifts. In particular, these authors draw upon governmentality literature, citing Dean (1999), Rose (1996), Miller and Rose (1990), in exploring the state's attempts to shape conduct of individuals, through establishing truths and norms. In analysing the shift towards marketisation in Australian agricultural industries, most notably the Australian wheat export market, I analyse the role performed by discourse in shaping knowledge, values, and identities, in making this transformation possible. Therefore, the relationship between discourse, knowledge, and power is integral to this work.

The poststructuralist approach to discourse analysis is influenced by the work of Michel Foucault, and considers discourse as "part of a process through which things and identities get constructed" (Lees 2004, pp. 102–103). For Foucault, discourse includes the use of language, however also includes ideology, strategy, and the relationship between knowledge and power (Lees 2004, p. 103; Sharp and Richardson 2001, pp. 195–196; Hall 1997, p. 44). Drawing on Foucault, Oels (2005, pp. 189–190) describes discourses as being groups of statements, which exist in a constant state of flux and are not necessarily adherent to any particular ideology. According to Oels (2005, p. 190), discourses are not produced from any one source. Rather, discourses are developed from "different local centres of power relations, but they are linked up in complex ways to form an overall strategy" (Oels 2005, p. 190). In this regard, as Feindt and Oels (2005, p. 165) describe, "Power is understood as a web of force relations made up of local centres of power around which specific discourses, strategies of power and techniques for the appropriation of knowledge cluster." From this perspective, discourse does not simply reflect power relations; discourse performs an active, integral role shaping power relations, and in turn, actively constitutes and organises society (Jacobs 2006; Atkinson 1999, p. 60; Hall 1997, p. 44; Laclau and Mouffe 1985, p. 96, cited in Howarth 2010, p. 311).

Discourse determines the structure of debate by developing "regimes of truth", which act to impose boundaries around how a problem can be defined, while also limiting the potential solutions to this problem (Guizzo and de Lima 2015; Jacobs 2006; Hall 1997). In this sense, discourses actively structure what can be considered as truth, and what cannot, while also directing how people should and should not act (Anderson 2010; Stenson and Watt 1999; Atkinson 1999). Discourse therefore shapes how people make sense of the world, by legitimising and delegitimising practices, values, and knowledges (Stenson and Watt 1999). As mentioned by Feindt and Oels (2005, p. 164), "A discourse constitutes specific ways of being engaged with the world and of being related to it." Thus, discourse performs a role in shaping reality or, at least, seeking to shape what people experience as reality (Feindt and Oels 2005, p. 164). As Rose (1993, p. 289) suggests, language should therefore be considered as the means through which the world can be understood and operationalised, in ways that can be acted upon by those who might seek to govern, such as politicians or experts, and also by those who occupy those domains which are recast in this light, such as the market and the family. Rose (1993, p. 289) states that "political discourse is more than ideology or rhetoric. It should be seen, rather, as a kind of intellectual machinery or apparatus for rendering reality thinkable in such a way that it is amenable to political deliberations." In this work, I draw on this interpretation to explore first how the Australian wheat industry and its farmers have been rationalised, made calculable and therefore knowable, and how this is then acted upon, through governmental technologies.

In turn, discourse and the creation of truth regimes affect how we recognise, understand, and solve problems. This is a considerable source of power, particularly in relation to policy making, where the rules governing what can be known, how this can be known, and what knowledge can contribute to formulating a legitimate policy response are implicitly shaped by discourse. Discourse, in this sense, is not only a product of power; it also has a reciprocal effect on the production of power and the legitimation of power (Oels 2005). Texts, in this regard, are integral in producing, reproducing, and legitimising truths, which on the one hand seek to reinforce the dominant discourse and on the other aim to further marginalise alternate discourses (Jacobs 2006; Feindt and Oels 2005; Stenson and Watt 1999). As Feindt and Oels (2005, p. 164) explain, "By delineating legitimate forms of truth production from illegitimate ones, a discursive formation includes the establishment of the terms of its reproduction and the allocation of empowering and disempowering subject positions." As a result of this active process of truth-making, and the fluidity of discourses, discourse is a site of contestation (Jacobs 2006; Feindt and Oels 2005).

Significantly, neoliberalism seeks to define itself as being apolitical, which implies that the "truth", as constructed by neoliberal discourses, is beyond reproach due to its substantiation by disinterested experts (Peck and Tickell 2002, p. 400). This truth, consisting of liberalised markets, self-interested and rational individuals, and constructs of competition, efficiency, and productivity, is conceptualised not as an ideology, or as a way of understanding and acting on society. Rather, this truth is portrayed as being common sense, to which there is no alternative (Dean 1999). In this regard, competing discourses challenging these constructs and ways of understanding are portrayed as lacking sound thinking and logic. In this manner, dominant discourses "create a series of absent agendas, agents, objects of concern and counter-narratives, which are mobilised out of the discursive picture" (Stenson and Watt 1999, p. 192). Discourse, from this perspective, holds considerable power. The dominant discourse—for example, neoliberalism—directs "thought and action" to align with thoughts and actions which can be considered legitimate within that neoliberal discourse (Hall 1997, p. 44). This approach to discourse analysis guides the development of the methodology for this research, which aims to understand how ideas and constructs which accorded with the neoliberal discourse came to be understood in general debate as "truths", while alternate discourses were essentially rendered as insignificant or without basis. The outcome of this contest, in this regard, is central in shaping how knowledge and truth is constructed, and how power is distributed. In turn, this affects how reality is constituted and made "thinkable" and operable (Rose 1993, p. 289; Miller and Rose 2008).

As Feindt and Oels (2005, p. 169) suggest, this also creates an important role for discourse analysis, in identifying how competing discourses have become marginalised. According to Feindt and Oels (2005, p. 169), writing in relation to environmental discourses:

> ...discourse analysis offers a reflexive understanding of 'the political' and transforms the practice of policy analysis. Discourse analysis allows one to study the power effects produced by and built into environmental discourse. The environmental discourse that constitutes an environmental problem enables and constrains the available policy options and the range of legitimate actors for its resolution. Discourse analysis can draw attention to marginalised discourses which offer alternative policy options.

This helps illustrate how I apply discourse analysis in relation the Australian agricultural restructuring, which not only explores how neoliberal discourses have created truth regimes, centred on efficiency, the individual, and the market however also analyses how the social world has effectively been erased in this process.

According to Rose (1993, p. 289), related to the use of discourse as a technology of governance is the conceptualisation of the "authority of truth", which Rose describes as being enabled by the role of the expert (Rose 1993, p. 297). This conceptualisation is integral to the relationship between power and knowledge, whereby the truth is systematically produced by "expert, authoritative or scientific discourse" (Anderson 2010, p. 50). As I analyse in the next section, quantification and calculation, reflecting knowledge produced by experts, such as neoclassical economists, is prioritised as expert knowledge, which has implicit consequences for what it is that can be known, and what information has credibility in policy making. In the case of Australian agriculture, as I suggest in this work, the prioritisation of the knowledge held and produced by particular types of experts has had significant ramifications for how the reality of agricultural restructuring has been constructed. In turn, this has considerable implications for how the Australian wheat export market has been problematised, and then acted on, by policy makers. In Australian agricultural policy discourses, it is the agricultural economist whose knowledge, and capacity to create knowledge through numbers, which is highly valued by policy makers as a form of expertise.

Sociology of Quantification

Related to governmentality research, I suggest, is literature exploring the sociology of quantification. In terms of making society calculable, and therefore able to be acted upon, quantification of the social world is critically important. This has the effect of shaping how the world can be understood, and shaping what matters in policy making. In addition, in relation to what can be considered as knowledge, the development of quantitative information as communicating incontrovertible "facts", as opposed to the claimed biases and limitations of subjective information, is a critically important shift in Australian policy making.

Quantification, described by Espeland and Stevens as "the production and communication of numbers", has become increasingly prominent in attempts to understand and act upon social phenomena (2008, p. 402). According to Espeland and Stevens (2008, p. 402), quantification can have numerous purposes and meanings and "Only by analysing particular instances of quantification in context can these purposes and meanings be revealed." I draw upon this interpretation to analyse how the Australian wheat industry, with all its complexities, has been narrowed to focus on quantitative measures such as wheat prices.

Quantification is an important part of this process. Quantification reduces complexity and removes the importance of context, making the objects of quantification more amenable to governing (Espeland and Stevens 1998). Numbers simplify society by reducing complex relations and situations to that which can be amenable to calculation (Friedberg 2013; Scott 1998; Espeland and Stevens 1998). Espeland and Stevens (1998, 2008) refer to this process as commensuration, where qualities are transformed into quantities. Scott (1998, p. 11) suggests this helps reduce phenomena to something which is "more legible and hence more susceptible to careful measurement and calculation". The reduction of complex phenomena to that which can be analysed, understood, and communicated in quantitative form helps to shape how problems can be identified, understood, and addressed (Espeland and Stevens 2008). In that sense, complex problems are reduced to solvable equations. Governing, in this instance, becomes a technical operation focused on the quantifiable performance of the economy.

Numbers are constructed as the product of disinterested, mundane technologies. Therefore, numbers are held as pure representations of knowledge and of reality (Miller and Rose 1990). Furthermore, whereas numbers can be used to create distance between governance regimes and the objects of governing, policy makers are able to distance themselves from the technologies used to generate numbers, and the numbers that are used to legitimise political power (Porter 1995; Rose 1991, 1993). Thus, numbers are claimed to be objective, neutral, and apolitical, not the political judgement of policy makers (Le Gales 2016; Pritchard 2005; Porter 1995; Miller and Rose 1990). Diaz-Bone and Didier (2016), Diaz-Bone (2016), and Le Gales (2016) challenge this claimed neutrality of numbers, arguing that the norms and conventions which are established to define the objects of measurement and how they are measured subtly influence how these numbers are produced. In that sense, numbers are not neutral, as is frequently implied by policy makers in Australia. Furthermore, the reality that they are often used to portray is not self-evident (Espeland and Stevens 2008). Despite this, advanced liberal governance regimes use numbers produced by quantification to represent a simplified, yet disinterested and unequivocal representation of reality, and act upon this knowledge to create policy which governs society (Espeland and Stevens 2008; Scott 1998).

Operationalising the Rationality of Markets, Firms, and Consumers

Technologies of Agency

As highlighted in the preceding section, quantification helps construct individuals who are calculable and who calculate their own performance accordingly. Technologies of agency help calculating individuals to improve their performance according to quantitative knowledge. In this regard, Dean (1999, pp. 167–169) argues that the construction of the self-governing individual is made possible through technologies of agency. Similar to technologies of empowerment, which aim to shape attitudes and behaviours under the guise of empowerment, technologies of agency encourage people to accept responsibility and become self-reliant through building and exercising agency (Herbert-Cheshire 2000; Larner 2000). This discursive construction reshapes the role of government to be a facilitator of skill development, an enabler which encourages active citizenship (McKee 2008, 2009; Herbert-Cheshire and Higgins 2004; Larner 2000; Rose 1996). This approach has been applied in an agricultural context, most fruitfully by Higgins (2001b, 2002a, b, 2005). In analysing the Agriculture—Advancing Australia package implemented by the Howard government in the late-1990s, Higgins argued that, rather than withdrawing assistance from agriculture, as is commonly believed, this package instead reshaped assistance to encourage farmers to adopt an economically rational approach to their farm practice (2001b, p. 324). This work develops the conceptualisation of farmers as self-reliant and individualistic. Higgins (2001b, 2002a, b, 2005) shows that, while the individualisation of farmers represents a shift away from state responsibility for farmers, this should be seen as a step towards the operationalisation of farmers, through the use of technologies of agency.

Responsibility, in this instance, is framed in terms of farmers' ability to develop skills and capacity to implement productivity enhancing measures. Since the 1970s, the state has carefully shifted the focus of its responsibility, from supporting farmers to remain on the land through policies providing various means of assistance, towards creating a policy environment which prioritises values such as productivity and efficiency, and facilitates and encourages farmers' entrepreneurialism and self-reliance. This book uses a genealogical analysis to make this shift visible. However, beyond this devolution of individual responsibility and risk towards individual farmers, I suggest that a further process is underway in Australia, which is shifting responsibility for farming towards private sector interests. As I argue, this has the potential to substantially undermine farmers' autonomy and transfer industry ownership and power towards financial investors.

Technologies of Performance

From exploring the shaping of individual conduct, I turn to the shaping of organisations' conduct—in particular, the AWB and farm lobby groups. Throughout the 1980s, policy makers sought to measure and assess the AWB's performance as a statutory marketer of wheat. The AWB's lack of objectives was framed as a problem, as this prevented governments from measuring its success in meeting these objectives. Thus, technologies of government, such as performance objectives (the maximisation of growers' returns), evaluation, audit, cost-benefit analysis, and econometric modelling, were introduced by policy makers to govern the AWB. Policy makers' attempt to construct performance objectives to reflect emerging policy truths of efficiency and cost-effectiveness is an important development in the deregulation process, which is under-developed in literature on agricultural deregulation in Australia. I suggest that the concept of governmentality provides a framework for understanding this shift, particularly through Dean's (1999) conception of technologies of performance. As described by Dean (1999, p. 169), "technologies of performance, then, are utilised from above, as an indirect means of regulating agencies, of transforming professionals into 'calculating individuals' within 'calculable spaces', subject to particular 'calculative regimes'." Therefore, whereas technologies such as auditing and benchmarking are ostensibly used to increase transparency and accountability, Dean (1999) and others such as Power (1996), Higgins et al. (2015), and Miller and Rose (1990) have suggested that these instruments are designed to coerce and control.

Quantification, coupled with technologies of government, is central to this subtle coercion. Quantification constructs a narrow and simplified interpretation of reality, grounded in the perception of numbers as disinterested and trustworthy (Desrosieres 2011; Rose 1993). The knowledge created through quantification is operationalised through assemblages of seemingly humble and routine technologies such as auditing, benchmarking, calculation, econometric modelling, data storage, and analysis, making this reality amenable to governing (Le Gales 2016; Rochford 2008; Kurunmäki and Miller 2006; Swyngedouw 2005; Pritchard 2005a, b; Larner 2000; Power 2000). Policy makers legitimise such technologies, claiming that they are essential in enhancing accountability, transparency, and credibility (Russell and Thomson 2009; Leander and Munster 2007). In this regard, technologies are constructed simply as examples of good governance practices. Governmentality theorists, on the other hand, argue that technologies of government are employed to define how phenomena can be understood and legitimise acceptable thought and action within this context (Rose 1991). As such, technologies of performance enable society to be governed "at a distance", through monitoring and controlling the minute actions of individuals and organisations (Penny 2016; Higgins et al. 2015; Russell and Frame 2013; Miller and Rose 2008; Higgins and Lockie 2002; Dean 1999). These technologies are used to govern behaviour, through establishing norms or expectations which legitimise certain behaviours and then monitor and assess the capacity of actors to perform according to these norms (Larner 2006; O'Malley et al. 1997; Rose 1993; Miller and Rose 1990). This could include the norms of the market, to which individuals, as moral, risk-averse, and economically rational actors, must conform (Larner 2006). In the case of wheat export marketing, this could include the construction of legitimate behaviour as the maximising of returns, and in relation to farm practice, the maximisation of productivity.

Governmentality scholars have analysed individual policies and policy collections through identifying assemblages of technologies such as calculation, audit, cost-benefit analysis, benchmarking, and performance objectives, to shape knowledge and actors and to act upon individuals and communities (Dufty 2015; McKee 2009; Rochford 2008; Lockie and Higgins 2007; Leander and Munster 2007; Kurunmäki and Miller 2006; Stratford 2006; Stenson and Watt 1999). For example, benchmarking and performance objectives act as a force for disciplining individuals, organisations, and nations and directing resources towards legitimised activities and behaviours (Sum and Jessop 2013; Connell 2013; Sum 2009; Rochford 2008; Leander and Munster 2007). As Sum and Jessop (2013, p. 37) state in relation to the Global Competitiveness Report and Global Competitiveness Index, which rank the economic performance of nations:

> These numerical scores and relative rank orders operate as a disciplinary tool (or paper panopticon) that draws more and more countries into its number order, comparing their economic performance scores over time and their ranking in relation to each other. [...] They deploy numbers and tables to rank countries. Annual revisions create a cyclical disciplinary art of country surveillance that institutionalises a continuous gaze through numbers that depicts countries' performance via changing rank and score orders.

Such disciplinary mechanisms govern behaviours, values, and attitudes by compelling actors to view their own purpose and performance in relation to the objectives which are established by the governing regime. In addition, benchmarking delineates between good and bad performance. Thus, the knowledge captured and communicated through this mechanism becomes central to interpreting what good performance is, how it can be measured, and, perhaps most pertinently, marginalising and delegitimising the activities, behaviours, and values which are not included within this knowledge.

Similarly, audit is a key technology used by government in this process, to facilitate legitimate behaviours (Miller and Rose 2008; Dean 1999). In "The Audit Society", Power (1997) illustrates audit as a mechanism for governing the conduct of organisations. Power (1996, p. 289) describes audit as a process of "making things auditable". According to Power, this process first involves the construction of a legitimised knowledge base, which is amenable to audit, and the development of environments which recognise and accept this knowledge base. Regarding public authorities, policy objectives, audit, and benchmarking become central to the governance of conduct (Rochford 2008; Power 1996, 2000). For example, in the case of wheat export marketing, this might include the use of audit to measure the AWB's ability to provide growers with premium prices, while reducing the cost of this activity to domestic consumers. Thus, if audit is being used to measure the AWB's performance against this objective, then the AWB is compelled to focus on improving its performance according to these measures.

Auditing presumes the existence of auditable "facts". Quantification of social phenomena creates these facts, through definition, measurement, and analysis. This is an important step in constructing what "matters" in policy terms. This construction entails creating objectives and benchmarks, and instruments for interpreting and assessing performance according to these standards (Diaz-Bone and Didier 2016; Desrosieres 2014, 2011; Rochford 2008). Quantification, and its reductionist, depersonalising approach, is central to this. As analysed by Rose (1991, p. 676), "To exercise power over events and processes distant from oneself it is necessary to turn them into traces that can be mobilised and accumulated." This simplifies the phenomena in question, creating distance between the governed and the governing, while enabling the state to monitor the performance and individuals and organisations, whose actions are guided by the performance measures which they are subject to. As quantification has developed in prominence, quantification of the social world has become established in many instances as a process for creating acceptable knowledge (Russell and Frame 2013; Russell and Thompson 2009; Scott 1998; Power 1996). It is rational, depersonalised, objective, and neutral (Pritchard 2005a). In that sense, the numbers created through quantification are held as having power unto themselves—they create a reality which is taken to be self-evident (Espeland and Stevens 2008). Governments are then able to use these numbers as a simplified, yet disinterested, unequivocal representation of reality, and act upon this knowledge to create policy which governs society (Friedberg 2013; Espeland and Stevens 2008; Scott 1998).

Muniesa, Millo, and Callon (2007, p. 4) describe mechanisms such as accounting methods and benchmarking procedures, economic modelling, and pricing techniques as abstractors. For Muniesa et al. (2007, p. 4), "'to abstract' is to transport into a formal, calculative space." This involves decontextualising objects, as part of the process of making goods calculable (Callon and Muniesa 2005). This concept of abstraction is particularly pertinent in relation to the deregulation of the wheat export market. On the one hand, I suggest that policy discourses have sought to marginalise the social function performed by the AWB, using benchmarking, performance objectives, and auditing to construct this organisation as having an exclusively economic role. Thus, the economic performance of the AWB is separated from its social function. In conjunction with this shift is the detachment of farming from its inherently social dimensions, including the attachment of farmers to their occupation, land, and identities as farmers and their role in supporting rural communities, economically and socially. Whereas quantification and technologies of performance prioritise the economic dimension of farming while excluding the social, technologies such as performance objectives, benchmarking, auditing, and economic modelling enable policy makers to verify the superiority of markets. These technologies not only render the wheat export industry as calculable, but also enable the construction of markets (Muniesa et al. 2007).

I suggest that in the case of the Australian wheat industry, policy makers have deployed assemblages of governmental technologies to facilitate the liberalisation of the Australian wheat export market. Studies examining policy through the governmentality lens have previously argued that the social has been reconfigured, with policy makers devolving risk and responsibility to individuals and their communities (Cheshire and Lawrence 2005; Tonts and Haslam-McKenzie 2005; Higgins 2002a; Stenson and Watt 1999). In the case of wheat export marketing policy in Australia, I suggest that rather, the social has been erased through technologies of performance which have sought to construct economic performance, as measured through wheat prices and supply chain costs, as being the only legitimate purpose of this policy area. In the policy debate around wheat export marketing, supply chain costs and wheat prices, for example, are portrayed by policy makers in precisely such ostensibly neutral numeric terms. As such, it is necessary to explore how these constructions have shaped wheat industry policy in Australia, and helped to facilitate the deregulation of the Australian wheat export market.

Genealogy of Wheat Industry Deregulation

This work examines the discursive constructions which made the shift towards private ownership of farming appear normal, desirable, and in farmers' best interests. To understand this change, we need to analyse the minute discursive shifts which shaped individual actors, policy values, and perceived responsibilities of the state towards farming. Thus, to answer the question of how deregulation was made possible, I have sought to construct a genealogy of wheat industry deregulation which captures these subtle shifts. In doing so, this research traces the discursive construction of truths, such as the good farmer, and the notion of farming as reducible to measures of efficiency and productivity, and therefore calculable and administrable. These truths are presented by discourses as neutral and normal. As with Higgins (2002a, p. 5), my use of genealogy is intended to investigate how these truths came to be established and to challenge the basis of these truths.

A genealogy of wheat industry deregulation is essential to understand the process by which this policy shift came to be viewed as a change that would benefit farmers, their communities, and the Australian society. Concepts such as competition, efficiency, individualism, technologies of audit and performance management, the passivity of firms in relation to the market, and the concept of the consumer as King (Jones 2012) are often presented in Australian policy discourses as self-evident truths, which therefore must be used to guide policy. Genealogical research using a poststructuralist approach to discourse analysis has enabled me to identify the subtle discursive shifts which have enabled these truths to become so ingrained in Australian policy making. In this research, I am seeking to understand the present policy environment by analysing previous discursive formations that have allowed this environment to develop as it has. As described by Kuchler and Linner (2012, p. 582), the purpose of genealogical research enables the researcher to "present a history of the present designed to outline the conflicts and strategies of control that condition discursive formations".

Each chapter in this book is underpinned by this genealogical approach to research, which aims to understand how Australian agricultural policy has developed to focus on economic indicators, while externalising social and environmental concerns. This book draws upon genealogical research that has analysed discourse, and its capacity to create knowledge through establishing rules governing what can be known, and how this knowledge can be understood (Van Herzele 2015; McMichael 2009; Jacobs 2006; Dixon and Hapke 2003; Higgins 2001b; Liepins and Bradshaw 1999). Building upon this work, I understand discourse as shaping policy debates through framing what legitimate knowledge is, and significantly, what legitimate knowledge is not. As mentioned by Higgins (2002b, p. 5), "genealogy provides a conceptually coherent means for challenging the search in historical investigations for the origins or foundations of knowledge," through exposing the "struggles for truth that underpin the claimed neutrality of these knowledges".

In this regard, genealogy is ideally suited to this research, which traces the construction of competition, efficiency, and markets as politically neutral truths, and analyses how these truths have facilitated policy change. Genealogical research examines processes of knowledge creation and power through discourse, while seeking to understand the silencing of alternate discourses (Van Herzele 2015; Kuchler and Linner 2012). This emphasis on the process of knowledge creation focuses genealogical research on understanding how truths are considered in the present, by analysing how truths have been constructed, and therefore, how they can be challenged (Hayter and Hegarty 2015). My genealogical research is inspired by Higgins' (2001b) article on the construction of the "low income farm problem". Higgins constructed a "genealogy of government", by exploring the problematisation of low-income farmers in the developing agricultural economics literature in the late 1960s and early 1970s.

In understanding the development of the "farm income problem", Higgins employed genealogy to "examine the discursive practices through which some rural producers were constituted, from the late 1960s, as a threat to the ongoing security of Australian agriculture" (Higgins 2001a, p. 359). Higgins sought to understand the influence of agricultural economists in not only defining the problem of "viability" but also providing the means "through which the conduct of farmers could be known, and acted upon" (Higgins 2001a, p. 359). This refers to the "conduct of conduct" that is characterised by the poststructuralist conception of governmentality, though also refers to the techniques of neoliberalism that were described by Rose (1993), such as auditing. Higgins (2001a, p. 359) also highlighted the role played by "experts"—in this case, agricultural economists—who sought to provide supposedly neutral appraisals of the problem of low-income farms and provide solutions which constructed the issue of "low-income farm problem" as one that could only be properly understood and addressed through agricultural economics (Higgins 2001a, p. 359). Higgins (2001a, pp. 361–362) refers to Rose and Miller, in stating that "expertise is central in translating the 'political' concerns of government – efficiency, industrial productivity, law and order, normality – into the politically 'neutral' discourse of management, social science, accounting and so forth." In that sense, the great strength of the employment of experts, or expert knowledge, in this regard, is that it is construed as being apolitical (2001, pp. 361–362).

Whereas Higgins' work analysed the problematisations of "unviable" farmers, and of government assistance, I extend this analysis by exploring the problematisation of statutory wheat marketing, and the construction of competition and efficiency as solutions to declining grower returns. By tracing developing policy discourses, I analyse how these ideas have become accepted as truths within wheat industry policy, and how this facilitated wheat export market deregulation. I have used this research to examine, not only policy itself, but also the process whereby policy is created and the discursive shifts which have influenced this process (Kuchler and Linner 2012). While my research considers key policy documents, such as Acts of Parliament or reports of government-initiated inquiries into the wheat industry, I also focus my research upon the "minor texts" such as newspaper articles or letters to the editor, which I consider as forming part of this discourse. These documents help provide context, though also help illustrate smaller shifts, which may not otherwise have been detected.

Documents

I have used this research to examine, not only policy itself, but also the process whereby policy is created and the discursive shifts which have influenced this process. While my research considers key policy documents, such as Acts of Parliament or reports of government-initiated inquiries into the wheat industry, I also focus my research upon the "minor texts" such as newspaper articles, which form part of this discourse. These documents, shown in Table 2.1, help provide context and help illustrate smaller shifts, which may not otherwise have been detected.

I focused my document searches at the State Library of Victoria, the National Library of Australia (online collections), and the Public Record Office of Victoria. I also searched the online databases of authorities such as the Productivity Commission and Australian Bureau of Agricultural and Resource Economics and Sciences (ABARES), as well as the websites of politicians, to locate speeches and media releases. I used the Australian Parliament House website to find Hansard transcripts, as well as information on government inquiries, including committee reports, submissions, and transcripts of public hearings. The newspaper articles published prior to 1994 were found in the Public Record Office of Victoria. Articles published between 1994 and 2004 were found in the State Library of Victoria microfilm collection, and articles published after 2005 were found using Google searches.

Table 2.1 Comprehensive selection of documents analysed in this work

Source type	*Author (year of publication)*
Reports prepared by consulting firms and independent think-tanks	ACIL Tasman (2004), Allen Consulting (2000a, b), Centre for International Economics (2005), ITS Global (2006), McKinsey Australia (2014), Nevile (1971)
Annual reports, 10-K forms, media releases, transcripts of speeches and presentations, published by agri-business corporations	ADM (2013, 2014, 2015a, b, c, d), Bunge Limited (2014a, b, 2015a, b), Conway (2012, 2013), GrainCorp (2012), Louis Dreyfus (2012, 2016), MacLennan (2014, 2015), Page (2014, 2015)
Newspaper articles (Sourced from the Public Record Office of Victoria (VA 1057, Grain Elevators Board, VPRS 9698, Historical Information and Reference Collection) and State Library of Victoria Microfiche collection)	The Age (1986, 1992, 1993), The Australian (1984, 1989, 1994, 1997), The Donald-Birchip Times (1992), Business Review Weekly (1992), Canberra Times (1992a, b), The Courier Mail (1989), Daily Commercial News (1992a, b, c), Financial Review (1984, 1986, 1992a, b), The Land (1986a, b, 1988, 1989, 1992), Stock Journal (1992), The Sun Herald (1992), Sydney Morning Herald (1992, 1996a, b), Telegraph Mirror (1993), The Weekend Australia (1986), 1989, 1992, The Weekly Times (1989, 1992)
Policy statements, speeches, interview transcripts, and media releases	Hawke (1986), Howard (1995a, b, 1996, 1997), Keating (1992, 1994), Kerin (1986), Turnbull (2015), Harper (2015), Hilmer (2013), Murdoch (2013), Sims (2015), Smith (1996)
Policies	Industry Assistance Act 1973 (Commonwealth), Rural Adjustment Act 1992 (Commonwealth), Wheat Marketing Amendment Act 2008 (Commonwealth), Wheat Marketing Amendment Act 2012 (Commonwealth). States Grants (Rural Adjustment) Act 1988 (Commonwealth)
Government publications	Commonwealth of Australia (1986, 1988a, b, 1992, 2004, 2008a, b, 2012, 2013, 2015, 2016)
Independent inquiries initiated by government	Harper, Anderson, McCluskey, and O'Bryan (2015), Hilmer, Rayner and Taperell (1993), Irving, Arney, and Linder (2000), Royal Commission into Grain Handling, Storage and Transport (1988)

(*continued*)

Table 2.1 (continued)

Source type	*Author (year of publication)*
Reports and papers by government authorities and departments	Australian National Audit Office (1989, 1990, 1991, 1992, 1993, 1994), Burdon (1993), Bureau of Agricultural Economics (1987), Department of Prime Minister and Cabinet (2014), Department of Social Services (2015, 2017), Essential Services Commission (2006), Industries Assistance Commission (1983, 1988, 1989), Industry Commission (1991, 1996), Jobs for New South Wales (2016), Knopke, O'Donnell, and Shepherd (2000), Malcolm, Davidson, and Vandenberg (2000), National Commission of Audit (2014), National Competition Council (2004, 2009), Nossal and Gooday (2009), Nossal, Zhao, Sheng, and Gunasekera (2009), Productivity Commission (2000a, b, 2001, 2005a, b, 2010, 2013, 2016, 2017), Rural Industries Research and Development Corporation (2007, 2011), Wheat Exports Australia (2012)
Papers by lobby groups	Business Council of Australia (2014), Joint Industry Submission Group (2000), National Farmer (1988), National Farmers Federation (1993, 2013), NSW Farmers (2014), Australian Wheat Board (1987)

I explore wheat industry deregulation as a process which occurred across four decades, from the 1970s to the present. In that sense, I aim to understand deregulation in the present, by exploring the discursive shifts which made this possible. These documents have been collected, as they share an interest in shaping economic, agricultural, and competition policy in Australia, and do so in a way that is relevant to agricultural restructuring and farming, more broadly. Some documents are integral to agricultural policy making, such as key reports resulting from government-initiated inquiries and Acts of Parliament. Some documents could be considered as "minor texts" (Rose, O'Malley, and Valverde 2006, p. 86), such as Newspaper articles featuring comments from politicians and other key stakeholders in agricultural policy, or media releases and speeches from key figures representing agricultural corporations, for example.

In analysing these documents, I examine how ideas such as efficiency, competition, markets, and the consumer are portrayed in these discourses, and how they are used to shape policy, by legitimising arguments favoured by policy makers and marginalising competing discourses. In relation to constructions of farmers, I analyse policy documents' portrayal of choice, freedom, self-reliance, and independence as goals for farmers to desire from policy, and to aspire towards exhibiting in their own practice. Moreover, I analyse constructions of firms as being the central actors in improving consumer well-being, in contrast with the problematisations of the state in performing this task. In analysing documentation representing farmers, farming, government organisations including the AWB, and also firms, I aim to identify important, yet subtle discursive shifts which contributed to this policy change.

Conclusion

Governmentality offers an important lens for understanding how the Australian wheat export industry came to be deregulated, supposedly in farmers' interests. Broadly speaking, I apply this approach to first identifying how the rationality of markets, firms, and individuals came to dominate Australian policy making. I argue that this was made possible through the construction of truth regimes, which posited that constructs such as efficiency and competition were essential ambitions of good policy, whereas previously held notions of collectivism and of society were portrayed as being redundant ideas belonging to a bygone era. Second, I use this approach, in conjunction with sociology of quantification, to analyse the construction of legitimate knowledge as being quantitative data, typically generated in markets, including costs and prices. Third, I explore the shaping of actors, in this case farmers and organisations, as sites through which governing occurs. This is made possible through technologies of government, specifically agency and performance, which are used to act on society. I employ a poststructuralist approach to discourse analysis, in conjunction with genealogical research, to identify the discursive shifts which have helped make these constructions possible. Drawing on documents supporting the past four-and-a-half decades of Australian agricultural and economic policy, I show how the deregulation of the Australian wheat industry has been made possible, and how this has contributed to the shift in power towards private investment and agricultural corporations.

References

ACIL Tasman, (2004), *A review of the NCP Grain Market Reforms: NCC occasional series*, Prepared by ACIL Tasman, Canberra: Ausinfo.

ADM, (2015a), *Form 10-K; Annual report pursuant to section 13 or 15(d) of the Securities Exchange Act of 1934, for the fiscal year ended December 31, 2013*, [online] http://www.sec.gov/Archives/edgar/data/7084/000000708414000011/adm-20131231x10k.htm [Accessed December 16 2015]

ADM, (2015b), *Archer Daniels Midland Co at Consumer Analyst Group of New York Conference*, February 17 2015, [online] http://www.adm.com/EN-US/NEWS/PRESENTATIONS/Pages/default.aspx [Accessed March 1 2016].

ADM, (2015c), *2014 Annual Report*, [online] http://www.adm.com/en-US/investors/Documents/2014-ADM-Annual-Report.pdf [Accessed March 1 2016].

ADM, (2015d), *ADM urges approval of Trans-Pacific Partnership*, [online] http://www.adm.com/news/_layouts/PressReleaseDetail.aspx?ID=691 [Accessed December 21 2015].

ADM, (2014), *ADM completes acquisition of remaining stake of Toepfer International*, [online] http://www.adm.com/news/_layouts/PressReleaseDetail.aspx?ID=589 [Accessed December 21 2015].

ADM, (2013), *Submission to the Senate Rural and Regional Affairs and Transport References Committee: The Ownership Arrangements for Grain Handling*, [online] http://www.aph.gov.au/DocumentStore.ashx?id=ca1c7a14-741b-4d6c-a4e6-d051dd054d84&subId=5640 [Accessed December 16 2015].

Allen Consulting, (2000a), *The Wheat Marketing Act 1989 – The Economic Impact of Competitive Restrictions*, Sydney: Allen Consulting Group.

Allen Consulting, (2000b), *The Wheat Marketing Act 1989 – The Social Impact of Competitive Restrictions*, Sydney: Allen Consulting Group.

The Age, (1993), *Look to marketing, processing, says farm leader*, May 12 1993.

The Age, (1992), *Kennett warns of public sector job cuts*, August 20 1992.

The Age, (1986), *Wheat blockade threat*, February 15 1986.

Alston, M., (2007), Globalisation, rural restructuring and health service delivery in Australia: Policy failure and the role of social work?, *Health and Social Care*, 15(3), pp. 195–202.

Alston, M., (2004), Who is Down on the Farm? Social aspects of Australian agriculture in the 21st century, *Agriculture and Human Values*, 21, pp. 37–46.

Anderson, A., (2010), *The constitution of youth: Toward a genealogy of the discourse and government of youth*, PhD Thesis, RMIT University.

Argent, N., (2005), The neoliberal seduction: Governing-at-a-distance, community development and the battle over financial services provision in Australia, *Geographical Research*, 43(1), pp. 29–39.

Atkinson, R., (1999), Discourses of partnership and empowerment in contemporary British urban regeneration, *Urban Studies*, 36(1), pp. 59–72.

The Australian, (1997), *Farm figures point to broader issues*, February 6 1997, State Library of Victoria Microfilm Collection.

The Australian, (1994), *Seeds of farm relief must be carefully sown*, September 21 1994, State Library of Victoria Microfilm Collection.

The Australian, (1989), *Ultimatum from wheat growers*, January 20 1989.

The Australian, (1984), *Editorial: Bringing together the two Australias*, August 28th 1984.

Australian National Audit Office, (1989), *Report on the audit of the Australian Wheat Board, 1988–1989*, Canberra: Australian Government Publishing Service.

Australian National Audit Office, (1990), *Report on the audit of the Australian Wheat Board, 1989–1990*, Canberra: Australian Government Publishing Service.

Australian National Audit Office, (1991), *Report on the audit of the Australian Wheat Board, 1990–1991*, Canberra: Australian Government Publishing Service.

Australian National Audit Office, (1992), *Report on the audit of the Australian Wheat Board, 1991–1992*, Canberra: Australian Government Publishing Service.

Australian National Audit Office, (1993), *Report on the audit of the Australian Wheat Board, 1992–1993*, Canberra: Australian Government Publishing Service.

Australian National Audit Office, (1994), *Report on the audit of the Australian Wheat Board, 1993–1994*, Canberra: Australian Government Publishing Service.

Australian Wheat Board, (1987), *Response to Industries Assistance Commission draft report of the wheat industry*, Melbourne: Australian Wheat Board.

Brasier, K., (2002), Ideology and discourse: Characterisations of the 1996 Farm Bill by agricultural interest groups, *Agriculture and Human Values*, 19, pp. 239–253.

Bunge Limited, (2015a), *Form 10-K; Annual report pursuant to section 13 or 15(d) of the Securities Exchange Act of 1934, for the fiscal year ended December 31, 2014*, [online] http://phx.corporate-ir.net/phoenix.zhtml?c=130024&p=irol-sec [Accessed March 7 2016].

Bunge Limited, (2015b), Unlocking Value: 2014 Annual Report, [online] http://phx.corporate-ir.net/phoenix.zhtml?c=130024&p=irol-reportsannual [Accessed March 7 2016].

Bunge Limited, (2014a), *Investor Day: Agribusiness and risk management overview*, October 12 2014, [online] http://phx.corporate-ir.net/phoenix.zhtml?c=130024&p=irol-presentations [Accessed March 7 2016].

Bunge Limited, (2014b), *Form 10-K; Annual report pursuant to section 13 or 15(d) of the Securities Exchange Act of 1934, for the fiscal year ended December 31, 2013*, [online] http://www.sec.gov/Archives/edgar/data/1144519/000104746914001600/a2218521z10-k.htm [Accessed December 16 2015].

Burdon, A., (1993), *Commonwealth Government Assistance for Adjustment in Agriculture*, Background Paper Number 14, Commonwealth of Australia, Canberra: Department of the Parliamentary Library.

Bureau of Agricultural Economics, (1987), *Wheat marketing and assistance*, Canberra: Australian Government Publishing Service.

Business Council of Australia, (2014), *Building Australia's comparative advantages*, [online] file://ntapprdfs01n01.rmit.internal/el4/E68334/Building_Australias_Comparative_Advantages_FINAL_28.7.2014.pdf [Accessed October 5 2016].

Business Review Weekly, (1992), *Farmers find a grain of comfort*, October 2nd 1992.

Canberra Times, (1992a), *Farmers may hold key to power*, September 28 1992.

Canberra Times, (1992b), *'Prove export monopoly superior to free market', Wheat Board dared*, September 30 1992.

Centre for International Economics, (2005), *Principles for national reform: Learning lessons from NCP*, Report prepared for the National Competition Council as part of the NCC Occasional Series, Canberra.

Cheshire, L., and Lawrence, G., (2005), Neoliberalism, Individualisation and Community: Regional Restructuring in Australia, *Social Identities*, 11(5), pp. 435–445.

Callon, M., and Muniesa, F., (2005), Economic markets as calculative collective devices, *Organisation Studies*, 26(8), pp. 1229–1250.

Commonwealth of Australia, (2016), *Smart farming: Inquiry into agricultural innovation*, report of the Standing Committee on Agriculture and Industry, [online] http://www.aph.gov.au/Parliamentary_Business/Committees/House/Agriculture_and_Industry/Agricultural_innovation/Report [Accessed December 20 2016].

Commonwealth of Australia, (2015), *Agricultural Competitiveness White Paper: Stronger farmers, stronger communities*, [online] http://agwhitepaper.agriculture.gov.au/ [Accessed June 6 2016].

Commonwealth of Australia, (2013), *National Food Plan, Our food future*, Department of Agriculture, Fisheries and Forestry, [online] http://www.agriculture.gov.au/style%20library/images/daff/__data/assets/pdf-file/0011/2293328/national-food-plan-white-paper.pdf [Accessed November 15 2016].

Commonwealth of Australia, (2012), *Parliamentary debates.* House of Representatives: official Hansard, Wednesday 31 October.

Commonwealth of Australia, (2008a), *Parliamentary debates.* House of Representatives: official Hansard, Thursday 13 March.

Commonwealth of Australia, (2008b), *Parliamentary debates.* Senate: official Hansard, Monday 16 June.

Commonwealth of Australia, (2004), *Review of the Agriculture – Advancing Australia package 2000–04*, Department of Agriculture, Fisheries and Forestry, [online] http://pandora.nla.gov.au/pan/41564/20040428-0000/www.affa.gov.au/corporate_docs/publications/pdf/industry_dev/aaa/aaareviewsummaryFeb2004.pdf, [Accessed November 15 2016].

Commonwealth of Australia, (1992), *Rural Adjustment Act 1992*, No.240 of 1992.

Commonwealth of Australia, (1988a), *States Grants (Rural Adjustment) Act 1988*, No. 112 of 1988.

Commonwealth of Australia, (1988b), *Australian wheat industry: Marketing in the future*, Canberra: Department of Primary Industries and Energy.

Commonwealth of Australia, (1986), *Economic and rural policy: A government policy statement*, Canberra: Government Printer.

Connell, R., (2013), The neoliberal cascade and education: An essay on the market agenda and its consequences, *Critical Studies in Education*, 54(2), pp. 99–112.

Conway, P., (2013), *Food security: The challenge ahead*, Speech presented at the MIT Food Symposium – Cambridge, Massachusetts, November 5 2013, [online] http://www.cargill.com/news/speeches-presentations/food-security-the-challenge-ahead/index.jsp [December 16 2015].

Conway, P., (2012), *One billion farmers can't be wrong: Why prices and markets matter*, February 8 2012, [online] http://www.cargill.com/news/speeches-presentations/one-billion-farmers-prices-markets-matter/index.jsp [Accessed March 29 2016].

Cooper, R., and Ellem, B., (2008), The neoliberal state, trade unions and collective bargaining in Australia, *British Journal of Industrial Relations*, 46(3), pp. 532–554.

The Courier Mail, (1989), *Grains Council division threatening new act*, December 20th 1989.

Daily Commercial News, (1992a), *Politics and the bane of NRC and rail reform*, September 28th 1992.

Daily Commercial News, (1992b), AGEA challenges AWB over single desk selling, October 2nd 1992.

Daily Commercial News, (1992c), AWB clash on wheat subsidies, September 10 1992.

Department of the Prime Minister and Cabinet, (2014), *Industry innovation and competitiveness agenda: An action plan for a stronger Australia*, [online] https://www.dpmc.gov.au/sites/default/files/publications/industry_innovation_competitiveness_agenda.pdf [Accessed October 5 2016].

Department of Social Services, (2017), *Welfare Reform: 2017 Budget*, [online] https://www.dss.gov.au/sites/default/files/documents/05_2017/budget_2017_-_welfare_reform_-_fact_sheet_for_web_0.pdf [Accessed May 30 2017].

Department of Social Services, (2015), *A new system for better employment and social outcomes*, Report of the Reference Group on Welfare Reform to the Minister for Social Services, [online] https://www.dss.gov.au/sites/default/files/documents/02_2015/dss001_14_final_report_access_2.pdf [Accessed May 30 2017].

Dean, M., (1999), *Governmentality: Power and Rule in Modern Society*, London: Sage Publications.

Dean, M., (1998), Risk, calculable and incalculable, *Soziale Welt*, 49, pp. 25–42.

Desrosieres, A., (2014), Statistics and social critique, *Partecipazione e Conflitto*, 7(2), pp. 348–359.

Desrosieres, A., (2011), The economics of convention and statistics: The paradox of origins, *Historical Social Research*, 36(4), pp. 64–81.

Diaz-Bone, R., (2016), Convention theory, classification and quantification, *Historical Social Research*, 41(2), pp. 48–71.

Diaz-Bone, R., and Didier, E., (2016), The sociology of quantification – Perspectives on an emerging field in the social sciences, *Historical Social Research*, 41(2), 7–26.

Dibden, J., Potter, C., and Cocklin, C., (2009), Contesting the neoliberal project for agriculture: Productivist and multifunctional trajectories in the European Union and Australia, *Journal of Rural Studies*, 25(3), pp. 299–308.

Dixon, D., and Hapke, H., (2003), Cultivating discourse: The social construction of agricultural legislation, *Annals of the Association of American Geographers*, 93(1), pp. 142–164.

The Donald-Birchip Times, (1992), *Grains Comment*, July 10 1992, Public Record Office of Victoria, VA 1057.

Dufty-Jones, R., (2015), Governmentalities of mobility: The role of housing in the governance of Australian rural mobilities, *Journal of Rural Studies*, 42, pp. 63–78.

Emery, S., (2015), Independence and individualism: Conflated values in farmer cooperation?, *Agriculture and Human Values*, 32, pp. 47–61.

Espeland, W., and Stevens, M., (2008), A sociology of quantification, *European Journal of Sociology*, 49(3), pp. 401–436.

Espeland, W., and Stevens, M., (1998), Commensuration as a social process, *Annual Review of Sociology*, 24, pp. 313–343.

Essential Services Commission, (2006), *Grain access review*, Issues paper, [online] http://www.esc.vic.gov.au/wp-content/uploads/archives/23015/1772_ESC_GrainIssuesPaper20060107.pdf [Accessed October 26 2016].

Feindt, P., and Oels, A., (2005), Does discourse matter? Discourse analysis in environmental policy making, *Journal of Environmental Policy & Planning*, 7(3), pp. 161–173.

Financial Review, (1992a), *The best in the bush*, August 28 1992.

Financial Review, (1992b), *Ideas Factor: Rural Revolution*, August 21st 1992.

Financial Review, (1986), *An open letter to wheatgrowers*, March 27 1986.

Financial Review, (1984), *Wheatgrowers demand abolition of government fees to end 'crisis'*, September 12 1984.

Fraser, C., Smith, K., Judd, F., Humphreys, J., Fragar, L., and Henderson, A., (2005), Farming and Mental Health Problems and Mental Illness, *International Journal of Social Psychiatry*, 51, pp. 340–349.

Friedberg, S., (2013), Calculating sustainability in supply chain capitalism, *Economy and Society*, 42(4), pp. 571–596.

Gibson, C., Dufty, R., Phillips, S., and Smith, H., (2008), Counter-geographies: The campaign against rationalisation of agricultural research stations in New South Wales, Australia, *Journal of Rural Studies*, 24, pp. 351–366.

Gray, I., and Lawrence, G., (1996), Predictors of stress among Australian farmers, *Australian Journal of Social Issues*, 31(2), pp. 173–189.

Gray, I., and Lawrence, G., (2001), *A future for regional Australia: Escaping global misfortune*, Cambridge: Cambridge University Press.

GrainCorp, (2012), *Media release: Proposal from Archer Daniels Midland Company (ADM) materially undervalues GrainCorp*, November 15 2012, [online] http://www.graincorp.com.au/_literature_113768/GrainCorp_response_to_ADM_conditional_proposal [Accessed June 5 2017].

Guizzo, D., and de Lima, I., (2015), Foucault's contributions for understanding power relations in British classical political economy, EconomiA, 16(2), pp. 194–205.

Hall, S., (1997), *The work of representation*, in Hall, S., (Ed.), Representation: Cultural representations and signifying practices, pp. 133–174, Sage, London.

Harper, I., (2015), *Speech to CEDA*, [online] http://competitionpolicyreview.gov.au/2015/04/14/speech20150414/ [Accessed July 14 2015].

Harper, I., Anderson, P., McCluskey, S., and O'Bryan, M., (2015), *Competition Policy Review: Final report*, [online] http://competitionpolicyreview.gov.au/final-report/ [Accessed July 30 2015].

Hawke, B., (1986), *A message from the Prime Minister*, The Mail-Times, February 28 1986.

Hayter, D, and Hegarty, P., (2015), A genealogy of postmodern subjects: Discourse analysis and late capitalism, *Theory and Psychology*, 25(3), pp. 369–387.

Herbert-Cheshire, L., and Higgins, V., (2004), From risky to responsible: expert knowledge and the governing of community-led rural development, *Journal of Rural Studies*, 20, pp. 289–302.

Herbert-Cheshire, L., (2000), Contemporary strategies for rural community development in Australia: a governmentality perspective, *Rural Society*, 16, pp. 201–215.

Higgins, V., (2002), *Constructing reform: Economic expertise and the governing of agricultural change in Australia*, New York: Nova Science Publishers.

Higgins, V., Dibden, J., and Cocklin, C., (2015), Private agri-food governance and greenhouse gas abatement: Constructing a corporate carbon economy, *Geoforum*, 66, pp. 75–84.

Higgins, V., (2005), *Governing agriculture through the managerial capacities of farmers*, in: Higgins, V., and Lawrence, G., (eds), Agricultural Governance: Globalisation and the new politics of regulation, pp. 118–132, London: Routledge.

Higgins, V., (2002a), *Constructing reform: Economic expertise and the governing of agricultural change in Australia*, New York: Nova Science Publishers.

Higgins, V., (2002b), Self-reliant citizens and targeted populations: The case of Australian agriculture in the 1990s, *Arena Journal*, 19, pp. 161–177.

Higgins, V., (2001a), Assembling restructuring: governmentality, economic regulation and the historical emergence of the 'enterprising farmer' in Australian agricultural policy, *Review of International Political Economy*, 8(2), pp. 311–328.

Higgins, V., (2001b), Governing the boundaries of viability: Economic expertise and the production of the 'low income farm problem' in Australia, *Sociologia Ruralis*, 41(3), pp. 358–375.

Higgins, V., and Lockie, S., (2002), Re-discovering the social: neo-liberalism and hybrid practices of governing in rural natural resource management, *Journal of Rural Studies*, 18, pp. 419–428.

Hilmer, F., (2013), *National Competition Policy: Coming of age*, Annual Baxt Lecture in Competition Law, presented September 19, 2013, [online] https://www.law.unimelb.edu.au/files/dmfile/130919-Sp-AnnualBaxtLectureIn-CompetitionLaw-FGH12Sept20132.pdf [Accessed July 30 2015].

Hilmer, F, Rayner, M., and Taperell, G., (1993), *National Competition Policy Review*, [online] http://ncp.ncc.gov.au/docs/National%20Competition%20Policy%20Review%20report,%20The%20Hilmer%20Report,%20August%201993.pdf [Accessed July 30 2015].

Howard, J., (1997), *Agriculture – Advancing Australia*, Media Release, September 15 1997.

Howard, J., (1996), The Liberal Tradition: The beliefs and values which guide the Federal Government, 1996 Sir Robert Menzies Lecture – presented November 18, 1996, [online] http://pmtranscripts.dpmc.gov.au/browse.php?did=10171 [Accessed July 30 2015].

Howard, J., (1995a), *Fair Australia: John Howard Headland Speech*, presented to the Australian Council of Social Service, October 13, 1995, [online] http://australianpolitics.com/1995/10/13/fair-australia-howard-headland-speech.html [Accessed July 30 2015].

Howard, J., (1995b), *The role of government*, Speech presented at The Menzies Research Centre, June 6 1995, [online] http://australianpolitics.com/1995/06/06/john-howard-headland-speech-role-of-govt.html [Accessed August 29 2016].

Howarth, D., (2010), Power, discourse and policy: Articulating a hegemony approach to critical discourse studies, *Critical Policy Studies*, 2(3–4), pp. 309–335.

Industry Assistance Commission, (1989), *Measuring the costs of government regulations: An initial exploration of the economy-wide effects*, Canberra: Industries Assistance Commission.

Industry Assistance Commission, (1988), *The Wheat Industry*, Report No. 411, AGPS, Canberra.

Industry Assistance Commission, (1983), *The Wheat Industry*, Report No. 329, AGPS, Canberra.

Industry Commission, (1996), *Industry Commission submission to the mid-term review of the Rural Adjustment Scheme*, [online] http://www.pc.gov.au/research/supporting/rural-adjustment-scheme/rural-adjustment-scheme.pdf [Accessed September 5 2016].

Industry Commission, (1991), *Statutory marketing arrangements for primary products*, Report No. 10, Canberra: Australian Government Publishing Service.

Industries Assistance Act 1973 (Commonwealth)

Irving, M., Arney, J., and Linder, B., (2000), *National Competition Policy review of the Wheat Marketing Act 1989*, Canberra: National Competition Policy – Wheat Marketing Act Review Committee.

ITS Global, (2006), *Sustaining Australia's Wheat Export Markets*, [online] https://www.awb.com.au/NR/rdonlyres/F41B17F4-B8D3-4467-9E02-AADEC063FDD9/0/ITS_Global_AWB_Report.pdf [Accessed October 26 2016].

Jacobs, K., (2006), Discourse analysis and its utility for urban policy research, *Urban Policy and Research*, 24(1), pp. 39–52.

Jobs for New South Wales, (2016), *Jobs for the future: Adding 1 million rewarding jobs for NSW by 2036*, [online] http://www.jobsfornsw.com.au/__data/assets/pdf_file/0020/90740/Jobs-for-the-future-full-report-August-2016.pdf [Accessed May 30 2017].

Joint Industry Submission Group, (2000), *Australian wheat: It's time for choice*, Submission to the Independent Review Committee reviewing the Wheat Marketing Act 1989 under National Competition Policy, Red Hill: Joint Industry Submission Group.

Jones, E., (2012), The market mechanism in reality and myth: Corporate business overlordship of small business, *Journal of Political Economy*, 68, pp. 62–85.

Keating, P., (1994), *Working nation: policies and programs*, Canberra: Australian Government Publishing Service.

Keating, P., (1992), *One nation: Statement by the Prime Minister*, Canberra: Australian Government Publishing Service.

Kerin, R., (1986), *Poor prospects ahead*, The Mail-Times, February 28 1986.

Knopke, P., O'Donnell, V. and Shepherd, A., (2000), *Productivity growth in the Australian grains industry,* ABARE Research Report 2000.1, Canberra.

Kuchler, M., and Linner, B., (2012), Challenging the food vs. fuel dilemma: Genealogical analysis of the biofuel discourse pursued by international organisations, *Food Policy,* 37(5): 581–588.

Kuehne, G., (2012), My Decision to Sell the Family Farm, *Agriculture and Human Values* 30(2), pp. 1–11.

Kurunmäki, L., and Miller, P., (2006), Modernising government: The calculating self, hybridisation and performance measurement, *Financial Accountability & Management,* 22(1), pp. 87–106.

Laclau, E., and Mouffe, C., (1985), *Hegemony and socialist strategy,* London: Verso.

Lees, L., (2004), Urban geography: Discourse analysis and urban research, *Progress in Human Geography,* 28(1), pp. 101–107.

Liepins, R., (1996), Reading agricultural power: Media as sites and processes in the construction of meaning, *New Zealand Geographer,* 52(2), pp. 3–10.

The Land, (1992), *Government shifts stance to 'good farmers',* August 27, 1992.

The Land, (1989), *Wheat has to 'get organised or get out',* January 19 1989.

The Land, (1988), *Growers say leave AWB as sole exporter,* June 30 1988.

The Land, (1986a), *Government wants farmers to survive and prosper: Hawke,* March 13 1986.

The Land, (1986b), *Beware of violence,* Editorial, March 13th 1986.

Larner, W., (2006), *Neoliberalism: Policy, ideology and governmentality,* in de Goede, M. (ed.), International political economy and poststructural politics, pp. 199–218, United Kingdom: Palgrave Macmillan.

Larner, W., (2000), Post-welfarist state governance: Towards a code of social and family responsibility, *Social Politics* 7(2): 244–265.

Leander, A., and Munster, R., (2007), Private security contractors in the debate about Darfur: Reflecting and reinforcing neoliberal governmentality, *International Relations,* 21(2), pp. 201–216.

Le Gales, P., (2016), Performance measurement as a policy instrument, *Policy Studies,* 37(6), pp. 508–520.

Liepins R, and Bradshaw, B., (1999), Neo-liberal agricultural discourse in New Zealand: Economy, culture and politics linked, *Sociologia Ruralis* 39(4), pp. 563–582.

Lockie, S., (2009), Responsibility and agency within alternative food networks: Assembling the "citizen consumer", *Agriculture and Human Values,* 26, pp. 193–201.

Lockie S, and Higgins, V., (2007), Roll-out neoliberalism and hybrid practices of regulation in Australian agri-environmental governance, *Journal of Rural Studies,* 23(1), pp. 1–11.

Louis Dreyfus, (2016), *What we do,* [online] http://www.ldcom.com/global/en/about-us/what-we-do/ [Accessed March 8 2016].

Louis Dreyfus, (2012), *2011 Annual Report*, [online] http://www.ldcommodities.com.br/Sobre_LDC/documentos/20120423_Annual%20Report_English%20Version.pdf [Accessed March 15 2016].

MacLennan, D., (2015), *Sustainability is the new normal*, April 21 2015, [online] http://www.cargill.com/news/sustainability-is-the-new-normal/index.jsp [Accessed March 29 2016].

MacLennan, D., (2014), *Hope for our planet: How a resilient global food system can feed a growing world*, Rabobank Duisenberg Lecture, Presented on October 11, 2014, [online] http://www.cargill.com/news/speeches-presentations/hope-for-our-planet/index.jsp [Accessed December 16 2015].

Malcolm, B., Davidson, B., and Vandenberg, E., (2000), *The Rural Adjustment Scheme: It's role, operation and effectiveness*, A report for the Rural Industries Research and Development Corporation, RIRDC Publication no. 99/19, Canberra; RIRDC.

McKee, K., (2009), Regulating Scotland's social landlords: Localised resistance to technologies of performance management, *Housing Studies*, 24(2), pp. 155–171.

McKee, K., (2008), Transforming Scotland's public-sector housing through community ownership: The reterritorialisation of housing governance, *Space and Polity*, 12(2), pp. 183–196.

McKinsey Australia, (2014), *Compete to prosper: Improving Australia's global competitiveness*, [online] http://www.mckinsey.com/~/media/mckinsey%20offices/australia/pdfs/compete_to_prosper.ashx. [Accessed August 10 2016].

McManus, P., Walmsley, J., Argent, N., Baum, S., Bourke, L., Martin, J., Pritchard, B., and Sorenson, T., (2012), Rural community and rural resilience: What is important to farmers in keeping their country towns alive?, *Journal of Rural Studies*, 28, pp. 20–29.

McMichael, P., (2009), A food regime genealogy, *The Journal of Peasant Studies*, 36(1), pp. 136–169.

Miller, P., and Rose, N., (2008), *Governing the present: Administering economic, social and personal life*, Cambridge: Polity.

Miller, P., and Rose, N., (1990), Governing economic life, *Economy and Society*, 19(1), pp. 1–31.

Mitchell, T., (2008), Rethinking economy, *Geoforum*, 39, pp. 1116–1121.

Muniesa, F., Millo, Y., and Callon, M., (2007) An introduction to market devices, *The Sociological Review*, 55(2), pp. 1–12.

Murdoch, R., (2013), *Free market is a fair market: Murdoch*, Speech presented to the Institute of Public Affairs, April 4th 2013, The Australian, [online] http://www.theaustralian.com.au/business/media/push-to-regulate-pressill-conceivedmurdoch/news-story/70177400a9b3aaf859a45bd96049873d [Accessed October 31 2017].

National Commission of Audit, (2014), *Towards responsible government: The report of the National Commission of Audit*, Phase One, Commonwealth of Australia, Canberra, [online] http://www.ncoa.gov.au/REPORT/docs/phase_one_report.pdf [Accessed October 5 2016].

National Competition Council, (2009), *Submission to Productivity Commission Inquiry into Wheat Export Marketing Arrangements*, [online] http://ncc.gov.au/images/uploads/SubPCWEM-001.pdf [Accessed October 26 2016].

National Competition Council, (2004), *NCC Occasional Series: A review of the NCP grain market reforms*, AusInfo, Canberra.

National Farmer, (1988), *National Farmer survey*, July 1 1988, Public Records Office Victoria, Melbourne.

National Farmers Federation, (2013), *Red tape in Australian agriculture: NFF Issues Paper*, [online] file://ntapprdfs01n01.rmit.internal/el4/E68334/NFF%20Issues%20Paper%20-%20Red%20Tape%20in%20Australian%20Agriculture%20Sept2013.pdf [Accessed December 20 2016].

National Farmers Federation, (1993), *New Horizons: A strategy for Australia's agrifood industries*, Canberra: National Farmers' Federation.

Nevile, J., (1971), *The government, the economy and productivity*, in Nevile, J., Le Couteur, G., and Bell, H., (eds), Productivity, Sydney: Committee for Economic Development of Australia.

Nossal, K., and Gooday, P., (2009), *Raising productivity growth in Australian agriculture*, ABARE project 3425, Canberra: Australian Bureau of Agricultural and Resource Economics.

Nossal, K., Zhao, S., Sheng, Y., and Gunasekera, D., (2009), Productivity movements in Australian agriculture, *Australian Commodities*, 16(1), pp. 206–216.

NSW Farmers, (2014), *Submission to Rural and Regional Affairs and Transport References Committee inquiry into Australian grain networks*, [online] http://www.aph.gov.au/DocumentStore.ashx?id=990367d0-7c89-4d83-9930-54eb96412e6e&subId=299362 [Accessed April 30 2015].

Oels, A., (2005), Rendering climate change governable: From biopower to advanced liberal government?, *Journal of Environmental Policy and Planning*, 7(3), pp. 185–207.

O'Keeffe, P., (2017a), Creating resilience or private sector dependence? Shifting constructions of the farmer in Australian policy discourses, *Space and Polity*, 21(3), pp. 318–334.

O'Keeffe, P., (2017b), Maximising efficiency, marginalising equity: A genealogy of Australian wheat export market deregulation and 'the good farmer', *Australian Geographer*, https://doi.org/10.1080/00049182.2017.1353585

O'Malley, P., Weir, L., and Shearing, C., (1997), Governmentality, criticism, politics, *Economy and Society*, 26(4), pp. 501–517.

Page, G., (2015), *Climate change and the future of food production*, October 12 2015, [online] http://www.cargill.com/news/speeches-presentations/climate-change-and-the-future-of-food-production/index.jsp [Accessed March 29 2016].

Page, G., (2014), *The price of responsibility*, Presented at the FT Commodities Summit, Switzerland, [online] http://www.cargill.com/news/speeches-presentations/the-price-of-responsibility/index.jsp [Accessed December 16 2015].

Peck, J., and Tickell, A., (2002), Neoliberalising space, *Antipode*, June 2002, pp. 380–404.

Penny, J., (2016) Between coercion and consent: the politics of "Cooperative Governance" at a time of "Austerity Localism" in London, *Urban Geography*, DOI: https://doi.org/10.1080/02723638.2016.1235932.

Potter, C., and Tilzey, M., (2005), Agricultural policy discourses in the European post-Fordist transition: Neoliberalism, neomercantilism and multifunctionality, *Progress in Human Geography*, 29(5), pp. 581–600.

Power, M., (2000), The audit society – second thoughts, *International Journal of Auditing*, 4, pp. 111–119.

Power, M., (1997), *The audit society: Rituals of verification*, New York: Oxford University Press.

Power, M., (1996), Making things auditable, *Accounting, Organisations and Society*, 21(2/3), pp. 289–315.

Pritchard, B., (2005), Unpacking the neoliberal approach to regional policy: A close reading of John Freebairn's 'Economic policy for rural and regional Australia', *Geographical Research*, 43(1), pp. 103–112.

Pritchard, B., (2005a), Implementing and maintaining neoliberal agriculture in Australia, *International Journal of Sociology of Agriculture and Food*, 13(2), pp. 1–14.

Pritchard, B., (2005b), Unpacking the neoliberal approach to regional policy: A close reading of John Freebairn's 'Economic policy for rural and regional Australia', *Geographical Research*, 43(1), pp. 103–112.

Porter, T., (1995), *Trust in numbers: The pursuit of objectivity in science and public life*, Princeton: Princeton University Press.

Productivity Commission, (2017), *Regulation of Australian agriculture*, Productivity Commission Inquiry Report, No. 79, [online] https://www.pc.gov.au/inquiries/completed/agriculture/report/agriculture.pdf [Accessed November 6 2017].

Productivity Commission, (2016), *Digital disruption: What do governments need to do?*, Commission Research Paper, Canberra.

Productivity Commission, (2013), *Productivity update*, May 2013, [online] file://ntapprdfs01n01.rmit.internal/el4/E68334/2013-pc-productivity-update.pdf [Accessed May 30 2013].

Productivity Commission, (2010), *Wheat export marketing arrangements: Report no. 51*, AGPS, Canberra.

Productivity Commission, (2005a), *Review of National Competition Policy reforms: Productivity Commission Inquiry Report*, No. 33, 28 February 2005, Commonwealth of Australia, Canberra.

Productivity Commission, (2005b), *Trends in Australian agriculture*, Canberra: Commonwealth of Australia.

Productivity Commission, (2001), *Structural Adjustment – Key policy issues, Commission Research Paper*, AusInfo, Canberra.

Productivity Commission, (2000a), *Single-desk marketing: Assessing the economic arguments*, [online] http://www.pc.gov.au/research/supporting/single-desk [Accessed December 20 2016].

Productivity Commission, (2000b), *Productivity Commission submission to National Competition Policy review of the Wheat Marketing Act 1989*, Melbourne: Productivity Commission.

Rochford, F., (2008), Sustainable rural water delivery: balancing resource and social sustainability, *The Australasian Journal of National Resources Law and Policy*, 12(1), pp. 59–85.

Rose, N., (1996), The death of the social? Re-figuring the territory of government, *Economy and Society*, 25(3): 327–356.

Rose, N., (1993), Government, authority and expertise in advanced liberalism, *Economy and Society*, 22(3), pp. 283–299.

Rose, N., (1991), Governing by numbers: Figuring out democracy, *Accounting Organisations and Society*, 16(7), pp. 673–692.

Rose, N., (1999), *Powers of freedom: Reframing political thought*, Cambridge: Cambridge University Press.

Rose, N., O'Malley, P., and Valverde, M., (2006), Governmentality, *Annual Review of Law and Social Science*, 2, pp. 83–104.

Royal Commission into Grain Storage, Handling and Transport, (1988), *Royal Commission into Grain Storage, Handling and Transport*, Canberra: AGPS.

Rural Industries Research and Development Corporation, (2011), *Agricultural productivity: Concepts, measurement and factors driving it: A perspective from the ABARES productivity analyses*, RIRDC publication No. 10/161, Canberra: Rural Industries Research and Development Corporation.

Rural Industries Research and Development Corporation, (2007), *Drivers of structural change in Australian agriculture*, RIRDC Publication No 07/057, Canberra: Rural Industries Research and Development Corporation.

Russell, S., and Frame, B., (2013), Technologies for sustainability: a governmentality perspective, *International Journal of Sustainable Development*, 16(1/2), pp. 91–106.

Russell, S., and Thomson, I., (2009), Analysing the role of sustainable development indicators in accounting for and constructing a Sustainable Scotland, *Accounting Forum*, 33, pp. 225–244.

Scott, J., (1998), *Seeing like a State: How certain schemes to improve the human condition have failed*, New Haven: Yale University Press.

Sartore, G., Kelly, B., Stain, H., Albrecht, G., and Higginbotham, N., (2008), Control, uncertainty, and expectations of the future: a qualitative study of the impact of drought on a rural Australian community, Rural and Remote Health, 8, p. 950.

Sharp, L., and Richardson, T., (2001), Reflections on Foucauldian discourse analysis in planning and environmental policy research, *Journal of Environmental Policy and Planning*, 3, pp. 193–209.

Sims, R., (2015), *Competition key to restoring Australia's productivity*, Speech delivered at the Infrastructure Partnerships Australia Conference, Sydney, September 4 2015 [online] https://www.accc.gov.au/speech/competition-key-to-restoring-australia%E2%80%99sproductivity [Accessed August 10 2016].

Smailes, P., (2000), The diverging geographies of social and business interaction patterns: a case study of rural South Australia, *Geographical Research*, 38 (2), pp. 158–181.

Smith, R., (1996), *Competition policy – implications of the Hilmer Report for the dairy industry*, Speech presented at the AIDC Annual General Meeting, December 4 1996, [online] https://www.accc.gov.au/system/files/Competition%20policy.doc [Accessed September 24 2015].

Stenson, K., and Watt, P., (1999), Governmentality and 'the death of the social'?: A discourse analysis of local government texts in South-East England, *Urban Studies*, 36(1), pp. 189–201.

Stock Journal (1992) *Editorial*, August 20th 1992.

Stratford, E., (2006), Technologies of agency and performance: Tasmania Together and the constitution of harmonious island identity, *Geoforum*, 37, pp. 273–286.

Sum, N., (2009), The production of hegemonic policy discourses: 'competitiveness' as a knowledge brand and its (re-)contextualisations, *Critical Policy Studies*, 3(2): 184–203.

Sum, N., and Jessop, B., (2013), Competitiveness, the knowledge-based economy and higher education, *Journal of the Knowledge Economy*, 4, pp. 24–44.

The Sun-Herald, (1992), *Country view: From the NSW Farmers Federation*, July 12 1992.

Swyngedouw, E., (2005), Governance innovation and the Citizen: The Janus Face of governance-beyond-the-State, *Urban Studies*, 42(11), pp. 1991–2005.

Sydney Morning Herald, (1996a), *High farm debt puts 100,000 in jeopardy*, July 4 1996.

Sydney Morning Herald, (1996b), *Farmers to blame, says scientist*, July 4 1996.
Sydney Morning Herald, (1992), *Little joy for farmers*, August 19 1992.
Telegraph-Mirror, (1993), *Farm gripe fires Keating ire*, March 10 1993, VA 1057, Grain Elevators Board, VPRS 9698, P1, unit 00015, Historical Information and Reference Collection.
Tonts, M., and Haslam-McKenzie, F., (2005), Neoliberalism and changing regional policy in Australia, *International Planning Studies*, 10(3), pp. 183–200.
Turnbull, M., (2015), *Transcript: Interview with Neil Mitchell on 3AW*, October 6 2015, [online] https://www.pm.gov.au/media/2015-10-06/interview-neil-mitchell-3aw [Accessed August 29 2016].
Van Herzele, A., (2015), A genealogy of urban forest discourse in Flanders, in Sandberg, L., Bardekjian, A., and Butt, S., (eds), *Urban forests, trees and greenspace: A political ecology perspective*, pp. 47–60, New York: Routledge.
Watts, R., (2016), *'Running on empty': Australia's neoliberal social security system, 1988–2015*, in Mays, J., Marston, G., and Tomlinson, J., Basic income in Australia and New Zealand: Perspectives from the neoliberal frontier, pp. 69–91, Hampshire: Palgrave Macmillan.
The Weekend Australian, (1992), *Survival of the fittest as dreams finally fade on land*, December 26–27 1992.
The Weekend Australian, (1989), *Farmers poised for $1m year of lobbying*, January 21st–22nd 1989.
The Weekend Australian, (1986), *Militant farmers to set political agenda*, February 15–16 1986.
The Weekly Times, (1992), *Drought plan furore*, August 26 1992.
The Weekly Times, (1989), *VFF gearing up for total deregulation of grain*, January 18 1989.
Wheat Marketing Amendment Act 2008 (Commonwealth).
Wheat Marketing Amendment Act 2012 (Commonwealth).
Wheat Exports Australia, (2012), *Submission to the Inquiry into the Wheat Export Marketing Amendment Bill 2012*, [online] http://www.aph.gov.au/parliamentary_business/committees/house_of_representatives_committees?url=arff/wheat/subs.htm [Accessed March 26 2014].

CHAPTER 3

Creating a Reality of Markets, Firms, and Consumers

Introduction

In this chapter, I examine the conceptualisation of the state, firms, and consumers in mainstream Australian policy discourses, as key actors in maximising the productive and efficient use of resources. According to this construction, these actors serve the broad, and rarely defined, purpose of enhancing living standards. I analyse these constructions to show how liberalised markets, and a "business-friendly environment", are framed as being beneficial for society, despite evidence of increasing inequalities arising from the neoliberalisation of society.

This chapter shows how efficiency is portrayed as a central ambition of policy making in Australia. Rather than developing policy to protect producers and workers, for example, the state's role has been recalibrated to focus on protecting the competitive process, in the belief that this will increase industry and market efficiency. Yet despite the pervasiveness of efficiency in Australian policy discourses, this term is often used vaguely, in the assumption that it is a good and worthy ambition. What efficiency represents, and what the implications of this policy ambition can be, is rarely articulated in any substantial detail in policy documents centred on competition, economic, and industry policy. This chapter interrogates these discourses to understand how efficiency is used in Australian policy making.

I show how efficiency is constructed as representing common-sense, logical thinking. This construction has the dual effect of reinforcing the

P. O'Keeffe, *Making Markets in Australian Agriculture*,
https://doi.org/10.1007/978-981-13-3519-8_3

perception, apparent in Australian policy making, that any person of sound mind will believe that efficiency is a good thing, while also marginalising competing discourses. In relation to this construction, competition, and the creation of winners and losers, is normalised as part of everyday life. In addition, economic restructuring and the marketisation of society are portrayed as the inevitable and necessary response to globalising processes, outside the control of the Australian government. The second part of this chapter draws upon constructions of firms and consumers, as essential actors in competitive, efficient markets. Mainstream Australian policy discourses claim competition in liberalised markets compels firms and producers to be efficient. In turn, this raises living standards for us as consumers, as competition between firms lowers the cost of goods and services, leading to better product quality and diversity as firms aim to appeal our needs.

I argue that prioritising the consumer focuses policy upon the end result—the price, quality, and choice of good or services. In turn, this undermines the importance of key actors and structures involved in the process of delivering products to consumers—the producers of the raw materials, the employees, and market concentration. I use this analysis to show how policy discourses shift attention away from the significance of market concentration and the power of large firms. Subsequently, policy makers tolerate consolidated markets as an outcome of this process, believing this market structure is the most efficient and best equipped to enhance consumers' living standards.

This has significant implications for farming, and specifically, the deregulation of the wheat export market. The process of maximising efficiency, and, subsequently, the market structure which achieves this aim, is framed as being "what matters" in policy terms. How commodities, goods, and services are produced is of lesser significance. As such, policy making has come to focus on creating favourable conditions for the actors which are perceived to maximise efficiency, such as large-scale firms. Those involved in the production process, such as workers and farmers, effectively become marginalised. The state perceives its role as being centred on the maximisation of productivity and efficiency in industries, through liberalised markets. In this context, consumers matter. Policy makers presume the benefits accruing to consumers filter through society, which is ultimately made more prosperous as a result. However, as I suggest in this chapter, the rising inequality within Australia suggests policy makers' faith in this logic is misplaced.

The State, the Market, and Society

Economic restructuring, which commenced in Australia throughout the 1980s in particular, reflected programmes of reform similar to those implemented by governments in the United States, United Kingdom, Canada, and New Zealand (Magnan 2015; Larner 2006; Le Heron and Roche 1999; Pusey 1996; Coleman and Skogstad 1995; Quiggin 1995). This shift was motivated by two key constructions. First, the construction of the state as the cause of economic malaise through its inefficiency and complacency and, second, the construction that liberalised markets and firms were motivated by self-interest and commercial disciplines, which were necessary to improve Australia's economic performance (IAC 1988).

Policy discourses problematised the state as the central cause of the apparently stagnating productivity and efficiency which was undermining Australian well-being (Fairbrother, Svenson, and Teicher 1997; Keating 1992; Mauldon and Schapper 1974). Examples commonly cited for state-driven stagnation include the imposition of tariffs, the use of statutory marketing arrangements, government underwriting for commodities prices, the assistance to industries such as agriculture and manufacturing, the regulation of labour markets and finance, and in some cases the restriction of foreign companies from entering markets (Pritchard 2005; Griggs 2002; Keating 1992; Martin 1990; Campbell 1974; Mauldon and Schapper 1974; Harris 1974). Policy makers agitating for change claimed that these policies restricted the Australian economy and its citizens by curtailing individuals' freedom, limiting choice, and undermining incentives to work hard and be innovative (Aulich and O'Flynn 2007; Keating 1992; Industry Commission 1991). Within Australia, the economic rationalism of the Hawke and Keating governments in the 1980s and early-to-mid-1990s focused on abruptly ending the state's intervention in markets and industries, essentially shifting control towards markets (Beer et al. 2016; Meagher and Wilson 2015; Pusey 1991; Gerritsen 1987). This included the corporatisation of federally owned airports (later privatised by the Howard government), the privatisation of the publicly owned Australian Airlines, Qantas and the Commonwealth Bank, and the deregulation of currency and financial markets (Beer et al. 2016; Aulich and O'Flynn 2007; Quiggin 1995, 2002; Fairbrother et al. 1997). In addition, reforms such as Working Nation and the Accord, developed by the Australian Council of Trade Union and the Labor government, sought to restructure unemployment, industrial relations, and policy relating to wage growth in Australia (Humphrys 2018; Jose and Burgess 2005; Stilwell 1994). The Howard government, which was elected in 1996, continued this policy trajectory through substantial reforms to healthcare, welfare, unemployment, and industrial relations policies (Morris and Wilson 2014; Wilson, Spies-Butcher, Stebbing, and St John 2013; Cooper and Ellem 2008; Aulich and O'Flynn 2007; McDonald and Marston 2005; Wilson and Turnbull 2005).

Policy makers claimed economic restructuring would result in a stronger, more prosperous economy. To some extent, this claim has come to fruition in Australia, which has enjoyed over two decades of uninterrupted economic growth, according to GDP figures. However, this growth in and of itself does not ensure that the benefits of economic prosperity are shared throughout society. The urgency of this question is evident, given that in recent decades wage growth has stagnated (Ryan 2015; Quiggin 2009; Stilwell 1991), and work has become increasingly precarious as secure, full-time employment becomes scarcer (Watts 2016; Healy 2016; Beer et al. 2016; Benach et al. 2014; Campbell 2008; Western et al. 2007). Shifts such as this have contributed to escalating economic inequalities within Australian society, which are most intensely felt by vulnerable members of the population, such as low-skilled workers, young people, and single mothers (Morley and Ablett 2016; Spies-Butcher 2014; Baird, Cooper, and Ellem 2009; Stilwell and Jordan 2007; Pusey 1998a, b). Despite the growing economic disparities evident in Australia, policy discourses maintain that markets are a great arbiter of fairness and therefore a strong economy will reward those who work hard, develop their skills, and use ingenuity (Harper et al. 2015; Murdoch 2013).

This policy shift relied on the underlying assumption that the strength of the Australian economy is central to the well-being of its citizens and of the broader Australian society. Driven by this assumption, policy makers presented economic growth and multifactor productivity as key indicators of economic performance, with the implication that strong performance against such measures reflects Australia's prosperity and the prosperity of Australian citizens. This argument, which has underpinned Australian competition and economic policy for the past three decades, suggests that liberalised markets, featuring competition between firms, deliver benefits to consumers. In this vision of the Australian economy, firms compete amongst each other by developing a better products and services, in a wider range of options and at a lower cost, to attract the custom of consumers. However, in policy discourses, the satisfaction and well-being of *consumers* is conflated with the well-being of *Australians*. As stated by

Harper et al. (2015, p. 23), in their review of Australia's National Competition Policy (NCP):

> Competition policy, like other arms of government policy, is aimed at securing the welfare of Australians. Broadly speaking, it covers government policies, laws and regulatory institutions whose purpose is to make the market economy better serve the long-term interests of Australian consumers.

In this regard, the welfare of Australians is directly related to their experience as consumers. Consumers, according to this report, are "not just retail consumers or households, but include businesses transacting with other businesses. In the realm of government services, consumers can be patients, welfare recipients, parents of school-age children or users of the national road network" (Harper et al. 2015, p. 31). According to this definition, there is no important distinction in competition policy discourses between people or business. What is important is the capacity of the market to function efficiently. The implication of this claim is that what is good for business is also good for consumers. The argument is underpinned by a number of assumptions, including the assumption that firms lack market power and are therefore best understood as being governed by the needs and wants of consumers. Business is framed here as a largely benevolent actor whose presence in social and economic life benefits consumers. The role of government is therefore free to shift from providing direct support to its citizens, towards creating a "business-friendly environment", which is assumed to be essential in attracting the business and private investment that is necessary to enhance the nation's economic performance through which citizens will then benefit as employees and consumers.

These assumptions, and policy makers' desire to create an economic environment which facilitates firms' access to previously restricted areas of Australian society, have created the conditions for corporate control of Agriculture. As Pusey (2016) has argued, this shift has been at the expense of Australian society, through this singular focus on attracting investment, increasing efficiency and productivity, and improving the contribution of agriculture to the national economy.

Why Efficiency?

Policy discourses frequently cite efficiency as a primary ambition of policy, though use this term vaguely, as if this were a general ambition to portray a sense of what policy should be achieving. The National Competition

Policy Review was initiated by the Keating government to investigate how a national policy could "develop an open, integrated domestic market for goods and services by removing unnecessary barriers to trade and competition" (Hilmer et al. 1993, p. 361). Introducing the findings of this review, Hilmer et al. (1993, p. 4) provide a rare, though brief, outline of what efficiency represents.

Hilmer et al. (1993, p. 4) draw upon three elements of economic efficiency: technical or productive efficiency, allocative efficiency, and dynamic efficiency. In particular, the concept of allocative efficiency is integral to the reorganisation of society and its resources. As outlined by Hilmer et al. (1993, p. 4), allocative efficiency is maximised when resources which are used in the production of good or services "are allocated to their highest valued uses". Competition is essential in enhancing allocative efficiency. As a process which creates winners and losers, competition compels firms to develop their capacity to use resources efficiently. The most successful, most efficient firms will succeed, whereas the firms which are least able to manage resources efficiently will fail. This process of competition is claimed to enhance allocative efficiency, as the resources held by the least efficient firms are freed-up and reallocated to the most efficient firms (Hilmer et al. 1993, p. 4). Hilmer et al. (1993) and the later Competition Policy Review, completed by Harper et al. (2015), each draw a connection between firm size and efficiency, favouring an environment where firm size and market share are not restricted, in the interests of maximising industry efficiency. Harper et al. (2015) use the term "efficient scale", to suggest that firms must reach a certain scale of operation before efficiency gains can be realised. Thus, allocative efficiency is achieved when the most efficient firms, presumed to be the largest in a market, are able to procure resources from inefficient firms, presumed to be the smallest, and use these resources to their productive potential (Harper et al. 2015; Hilmer et al. 1993).

Whereas policy discourses construct economic efficiency as an integral ambition of responsible government, firms are constructed as central actors in the process of maximising efficient resource use. Conversely, regulation which prevents firms' entry to markets and restricts firms' capacity to grow is portrayed in policy discourses as impairing the prosperity of Australian society. These constructions are significant. The state and firms are portrayed as having symbiotic relationship, where the needs and goals of each are intertwined. The facilitation of this relationship is portrayed as being in the interests of the broader society.

Efficiency as a Truth

Policy discourses foster the truths of efficiency, competition, markets, and consumers, through a series of discursive techniques aimed at buttressing the discourse of neoliberalism within Australian policy making. Marketisation, and the focus on facilitating the interactions between firms and consumers, is framed as an approach to economic organisation which has no legitimate alternative. In this context, policy makers' singular focus upon maximising the efficient use of resources becomes an inevitable, common-sense response to conditions which are portrayed as being beyond the control of individual governments.

The Depoliticisation of Markets

Policy discourses seek to normalise competition, efficiency, and markets, to the extent that they are framed as boring and apolitical (National Competition Council 2009; Howard 1995a, b). For example, in his speech to the Menzies Research Centre, titled "The Role of Government", then federal opposition leader John Howard (1995b) stated that "Sound money, responsible budgets and efficient markets are nothing more than mechanisms to deliver rising living standards." Howard (1995b) contends that while "it might be boring to many…any government interested in its country's economic future and jobs for its people must get on with the task of improving its efficiency." In addition, the reference to microeconomic reform as "boring" downplays the significance of Howard's statement. The task of improving efficiency is simply uninteresting; that is what good government should simply aim to do. There is limited public interest in this issue, it is implied, as there is no question that living standards are directly related to efficiency. Henceforth, there is no need for a public debate around the validity of this relationship.

Competition as a "Normal" Part of Life

The creation of winners and losers by competition is normalised as part of economic life (Department of Prime Minister and Cabinet 2014; Gray et al. 2014; National Farmers Federation 2013; Turnbull 2015; Productivity Commission 2001). This is described by the Productivity Commission (2001, p. 42):

> Individuals and firms win or lose from market-based changes every day. This is usually considered to be part of the normal operation of markets. For the losers, assistance beyond the social safety net and generally available measures is rarely provided.

This perspective contributes to the argument that the individual firms within a market should not be supported by competition policy, that instead it is the process of competition which should be protected. As Malcolm Turnbull said in a radio interview with conservative commentator Neil Mitchell following his accession to the role of Australian Prime Minister in 2015 (Turnbull 2015):

> Business is tough. Competition is tough. Once you start providing protections for one business after another you start putting up the price of products in Australia.

Turnbull's perspective reflects that of the man he deposed as Prime Minister, Tony Abbott, who explained "The guiding principle of the Government's Industry Innovation and Competitiveness Agenda is to focus on Australia's strengths, not to prop up poor performers" (Department of Prime Minister and Cabinet 2014, p. ii). This constructs competitive markets as necessarily tough environments, in which it is only the most efficient, most productive firms and industries that survive. Policy discourses normalise this process as an integral aspect of efficiency maximisation. Firms that fail in this process are portrayed as inefficient, incapable of meeting the needs of consumers, and undermine aggregate efficiency and productivity. Interference in this normal process, it is claimed, substantially undermines market efficiency.

Marketisation as Common Sense

The pervasiveness of the neoliberal discourse in Australian policy is emphasised in the National Commission of Audit (NCoA) report released in 2014, which was initiated by the Abbott government to focus on "the biggest and fastest growing areas of [Commonwealth] spending to identify savings" (NCoA 2014, p. i). To analyse the state's efficiency, the NCoA developed a 10-point framework, described as a "set of common sense principles to guide its deliberations", which it called the "Principles of Good Government" (NCoA 2014, pp. iii–iv). Point 9 states that responsible governments should:

> Act in the public interest and recognise the benefits of markets. In competitive markets, customers, not producers, take precedence. Competition and contestability drive lower costs, improve quality and give people what they want. Government should act in the public interest and only intervene in markets where market solutions fail to produce the best outcome for the nation as a whole. (NCoA 2014, p. iv)

According to this principle, marketisation of public services and industries is not an approach to government based on ideology, it is simply common sense; the actions of a responsible government. In addition, marketisation is claimed to be in "the public interest", as it gives people, as consumers, "what they want". What people want is therefore portrayed as being deliverable by competitive markets which prioritise consumers' needs. Producers, according to the NCoA, are relatively insignificant.

The "public interest" and "the nation as a whole" are also problematic terms. Measured in policy discourses through indicators such as multifactor productivity growth and GDP, the good of the nation is considered in terms of the health of the national economy. Thus, measures designed to improve economic performance according to these indicators are portrayed as justifiable on the basis that it is in the "national interest". This quells opposition to policy changes which may undermine the well-being of a section of Australian society, such as farmers. For example, policy makers' claim that wheat export market deregulation benefits consumers and enhances the well-being of "the nation as a whole" partly nullifies farmers' claims that this policy shift will negatively impact them. This creates an adversarial environment where farmers' interests are constructed as oppositional to the rest of Australian society. Furthermore, this constructs the support received by farmers through statutory marketing as a hindrance on the well-being of the broader Australian community. In addition, terms such as the "nation as a whole" and measures designed to reflect the nation's economic performance can be used to build a case for further economic restructuring.

The Inevitable Need to Maximise Competition and Efficiency

Policy discourses stress the urgency with which barriers to efficiency should be removed (McKinsey Australia 2014; Department of Prime Minister and Cabinet 2014; McKinsey Australia 2014; Hilmer et al. 1993). According to the arguments posited in these policy documents, unless efficiency is urgently pursued, living standards will decline. For example, in the foreword to the "Industry innovation and competitiveness" report, then Prime Minister Tony Abbott contended the urgency for further deregulation was heightened in response to the actions of other countries, stating that "Our competitors are cutting taxes, reducing compliance costs, building infrastructure and reining in government spending. We need do likewise if we are to compete with them." Conveying the urgency of these policy measures, the report constructs negative scenarios which attempt to sketch the predicted decline of Australian society in the case that these policy changes are not implemented (Department of Prime Minister and Cabinet 2014, p. iii).

The apparent urgency and inevitability of marketisation reconstructs questions around this policy shift as a matter of when, and how quickly, this change can occur, rather than whether this is actually in society's best interests. Ultimately, these constructions seek to create a reality in which firms operating in liberalised markets are portrayed as the key actors in raising consumers' living standards. Thus, the discursively constructed policy truths of efficiency, competition, and consumers, for example, are framed as unquestionable ambitions of good policy, while alternate discourses are dismissed as lacking credibility and antithetical to notions of advancement and prosperity. Whereas markets are portrayed as the central mechanism for enhancing living standards, the state, on the other hand, acts as a market steward, ensuring that markets are operating at maximum efficiency. This leads to the concept of the "business-friendly environment", as a construct for maximising society's prosperity.

Firms and Efficiency

Creating a Business-Friendly Environment

As stated by numerous policy documents agitating for further marketisation within Australian society, despite recent decades of economic restructuring, multifactor productivity growth has remained low in Australia for over a decade (DPMC 2014; Productivity Commission 2013, 2016; NCOA 2014). In recent years, key policy documents have attributed this economic malaise to the supposedly inefficient use of resources. The Productivity Commission, Department of Prime Minister and Cabinet, and National Committee of Audit each cite high production costs, particularly those associated with wages and employee entitlements as the key contributing factor which governments must address to improve the nation's economic performance (DPMC 2014; Productivity Commission 2013, 2016; NCOA 2014). The solution is to create a "business-friendly environment" and attract corporate investment by reducing regulatory restrictions and barriers to market entry, by lowering corporate taxes, and by curbing employee benefits (DPMC 2014, p. iii; Jobs for New South Wales 2016; NCOA 2014). Examples of this include the deregulation of the wheat export market, designed to facilitate private firms' entry into this industry and end the 60-year monopoly over wheat exports held by the AWB (O'Keeffe and Neave 2017).

The notion of the "business-friendly environment" as a solution to the nation's flat economic performance is based on the assumption that firms' increased investment in the Australian economy will stimulate economic activity and create jobs. In addition, as firms enter Australian markets, it is assumed that the rivalrous competition between these firms will compel greater resource efficiency. This is believed to then lift the productivity of Australian industries, and improve consumer welfare, as firms return the gains made through greater efficiencies to consumers. Thus, measures which are designed to benefit firms are portrayed in policy discourses as being in the broader interest of Australian society (DPMC 2014, p. xi). Whereas these benefits are captured in changes to national economic performance, such as employment statistics, multifactor productivity growth, and economic growth, policy discourses are silent on the effect of such shifts upon people who are directly impacted. Thus, the overall performance of the economy is prioritised over the measurable effects of the "business-friendly environment" upon vulnerable populations of workers, for example, who experience increasingly precarious work arrangements (Watts 2016; Healy 2016; Beer et al. 2016; Benach et al. 2014; Campbell 2008; Western et al. 2007), or the farmers who have exited agriculture following the deregulation of the wheat export market (Baker 2018; O'Keeffe and Neave 2017). The emphasis on economic growth and resource productivity, for example, presumes that these measures are effective indicators of societal health. Yet whereas there are groups who clearly suffer as a result of these shifts, such as retail and hospitality workers and farmers, the beneficiaries of enhanced economic growth and corporate investment are less obvious. The escalating inequality within Australia, observable since the increased marketisation and privatisation of society, suggests that the benefits of economic growth and prosperity are not being distributed evenly (Morley and Ablett 2016; Spies-Butcher 2014; Western et al. 2007; Stilwell and Jordan 2007; Pusey 1998a, b).

Nevertheless, the argument for the business-friendly environment persists, despite its many questionable assumptions. Firms, with their greater capacity to maximise efficiency and responsiveness to markets, are framed as the actors which will raise Australia's prosperity, shared or otherwise. Thus, policy discourses frame governments' central purpose as creating a business-friendly environment.

Shifting the Focus to Efficiency and Consumer Welfare

Having established efficiency, productivity, and the nation's economic performance as the primary ambitions of "good governance", policy discourses portray firms as the key actors in fulfilling these ambitions. Firms' performance, according to competition policy reviews by Hilmer et al. (1993) and Harper et al. (2015), is governed by competition in liberalised markets. Firms are compelled by commercial disciplines to maximise efficiency, thus minimising costs and enhancing productive resource use. Firms which are able to out-compete rivals in these key areas will therefore have greater appeal to consumers and succeed in this environment. This is claimed to enhance allocative efficiency, as the successful firms, presumed to be the most efficient, innovative, and productive firms, will subsume the resources and market share of the least successful firms as they exit the market. Thus, in this process, resources are reallocated to the most efficient firms, increasing industry efficiency and productivity. The reallocation of resources, through either firms' exiting the market, or strategies such as mergers and acquisitions, has the potential to reduce the number of market participants and create dominant participants in that market. However, policy discourses, evident in reports such as the Competition Policy Review (Harper et al. 2015), portray market structure, in terms of the number of firms in a given market, as being less important than the efficiency of that market.

This argument is grounded in economic and industrial organisation debates, which have sought to position the dominant firm as an actor which is integral in maximising efficient resource use. In his influential paper, "Industry structure, market rivalry, and public policy", Demsetz (1973) contends that firm efficiency determines profitability. According to Demsetz (1973), the most efficient, and therefore most profitable, firms expand market share. Thus, Demsetz (1973, p. 5) claims that, if concentration emerges due to the "superior efficiency of those firms that have become large, then a deconcentration policy, although it may reduce the ease of colluding, courts the danger of reducing efficiency either by the penalties that it places on innovative success or by the shift in output to smaller, higher cost firms that it brings about".

In other words, firms which are the most efficient, innovative, and responsive to consumers' needs will naturally grow their profitability and market share. Inhibiting this growth will potentially undermine industry and market efficiency, and reduces firms' incentives to grow their business. Critics such as Bork (1967, 1978), and Baumol and Ordover (1985, p. 247) argued the Federal Trade Commission, in restricting market concentration, punished firms that were successful due to their superior efficiency and economies of scale and scope.

According to Bork (1967, p. 242), focusing antitrust policy on addressing market concentration was:

> … creating a broad trend of policy directed less to the interest of consumers in free markets than to the interest of inefficient producers in safe markets. […] My thesis is that existing statutes can be legitimately interpreted only according to the canons of consumer welfare, defined as minimizing restrictions of output and permitting efficiency, however gained, to have its way.

Thus, the goals of antitrust, or competition policy, should be to maximise efficiency, or to use Bork's term, maximise consumer welfare, "however gained". Firms and market structure assume less significance in this conception. Efficiency, and the gains that this bestows upon consumers, becomes the central purpose of policy aimed at facilitating competition. Thus, attempts to reduce market concentration and curb the strength of dominant firms are portrayed not only as a punishment upon the most successful firms and a distortion of incentives, such actions are framed as harmful to allocative efficiency and, therefore, national prosperity. In turn, the development of large corporations with the capacity to exhibit substantial market power is excused, on the assumption that the presence of these firms increases efficiency, and firms, regardless of size, do not actually exert power over the market. These assumptions are evident in the definition of competition used by Hilmer et al. (1993) and Harper et al. (2015).

Consumers

The Technology of the Consumer

Policy discourses have constructed firms as actors which can raise national economic performance and therefore, national prosperity. Thus, the business-friendly environment, which caters to firms' needs, is portrayed as essential in enhancing societal well-being. In addition to this construction is the morphing of citizens to consumers in Australian policy. This is an essential element in the extension of markets throughout society. If people are constructed as consumers, then people's needs, well-being, and fulfilment are shaped as being best served by markets. This leads to the construction of "the consumer", as a governmental technology. Much of the governmentality-inspired literature focusing on consumers explores the operationalisation of consumers, for example, through ethical consumption (Powell 2018; Derkatch and Spoel 2017; Kremers and Brassett 2017; Guthman and Brown 2016). I suggest that the construct of "the consumer" can also be understood as a governmental technology, which helps to reorganise what matters in policy making, what has value, and helps us to understand how power has been shifted from "society" to "the market".

Within Australia, the construction of the citizen as consumer is implicitly connected to the privatisation programmes implemented initially by the Hawke-Keating Labor governments from 1983 to 1996, and then, the Liberal government under John Howard, which completed four parliamentary terms from 1996 to 2007 (Meagher and Wilson 2015; Aulich and O'Flynn 2007; Ryan 2005). Speaking as leader of the Liberal Party in 1995, Howard claimed modern governments must seek to "expand and enhance individual liberty, freedom, opportunity and choice, to help people help themselves" (Howard 1995a). Howard's emphasis on consumer choice is influenced by the notion that individuals are motivated by self-interest, which can be used in liberalised markets to make choices which maximise their utility (Stanford and Taylor 2013; Aulich and O'Flynn 2007; Ryan 2005). Governmentality research, in particular, has sought to analyse the construction of the individual, and associated rationalities of choice and freedom, as a means through which the state is able to govern through the rational, self-interested choices of individuals (Villadsen and Wahlberg 2015; Prince, Kearns and Craig 2006; Colebatch 2002; McDonald and Marston 2001). Associated with the apparent retraction of the state is the devolution of responsibility and risk to individuals. In this instance, the individual is portrayed as self-reliant and independent, and responsible for the consequences of their actions and choices. This leads to the construction of responsible choices, which are made according to the established norms, attitudes, and values, which are produced and reproduced in discourse (Scourfield 2007; Prince, Kearns, and

Craig 2006). Responsible choices, in this context, are constructed as economically rational choices, made by informed individuals who are capable of dispassionately assessing the costs and benefits associated with their decision making (Higgins and Lockie 2002).

Therefore, good choices are portrayed as rational, informed, and economically responsible choices made by consumers who are intent on maximising their utility. Policy discourses link choice with freedom, which is conceptualised as the freedom to choose. Albeit, a freedom which can be expressed within the boundaries of economic rationality and self-interest. In this construction of choice as a form of freedom, choice is not simply a means through which consumers are able to communicate their needs to producers, choice removes restrictions, opening up new possibilities and opportunities. Thus, choice is liberating. However, discourses in Australian policy, particularly around competition and economic policy, construct consumer choice as a mechanism for empowering people. As stated by Ian Harper (2015) in an address to the Committee for Economic Development of Australia (CEDA):

> ...competition in the markets for the goods and services we buy can empower us as consumers; empower us through an ability to choose. And when we can choose between providers, we see businesses competing with one another to attract our custom.

According to this conception of consumer choice, it is consumers who possess power in markets. Firms, on the other hand, simply respond to consumers' needs and desires, communicated through market signals. This constructs firms as having limited power to influence markets. Their success reflects their capacity to be efficient, innovative, and responsive to consumer sentiments. Consumer power, in this regard, is portrayed as the power to exercise choice between products and services produced by firms. Thus, marketisation is portrayed as a mechanism for giving people power over their lives, through providing a semblance of choice in relation to their roles as consumers. In this manner, policy discourses construct marketisation of public services, such as human services, for example, as a means for democratising society and empowering people to make choices. Choice is therefore granted a powerful status in policy making: for empowering people and for improving the efficient use of resources, as firms seek to improve their efficiencies, innovative capacities, and productivity to meet consumers' demands.

As consumers exercise choice, based on a dispassionate assessment of prices, value, and product quality, for example, consumers' choices determine which firms develop market share. Provided that consumers are able to identify the most successful firms, in terms of efficiency, innovation, and productivity, for example, this process will cause a reallocation of resources towards the best-performed firms. In turn, the efficiency and productivity of resource use will be enhanced. Thus, consumers' choices compel firms, in competitive markets, to become more efficient, which drives the economic performance of industries and, therefore, the nation's economic performance. The construction of efficient markets is therefore dependent upon consumers—consumers who make responsible, rational, and self-interested decisions. Importantly, this depends upon people's willingness to view themselves as consumers, and to view their power in terms of consumer choice. In addition, the extension of markets throughout society depends upon the notion that empowerment, freedom, prosperity, and well-being are best achieved through the efficient allocation of resources by liberated markets.

Raising Living Standards

Within Australia, policy makers claim that ultimately, competitive markets raise consumers' living standards. This argument is used repeatedly to justify the creation of a business-friendly environment. Neoliberal policy discourses have sought to construct the idea that enhancing living standards is achievable through liberalising markets and by extending markets into areas of social and economic life previously controlled by the state, such as human services and social services (Hilmer et al. 1993; Harper et al. 2015; Productivity Commission 2005a, 2016; McKinsey Australia 2014; DPMC 2014; NCOA 2014). As stated by the NCoA (2014, p. 9), "A market-based society with a meaningful role for government is a proven way of boosting living standards." Similarly, then Prime Minister Tony Abbott claimed increased "productivity and efficiency in markets" would "raise living standards for all Australians" (Abbott 2013). These statements, which are common in Australian policy discourses, imply that the connection between liberalised markets and raised living standards is so clear that there is no need to explain how market deregulation raises living standards. In addition, what "living standards" represent, and whose living standards are being raised through marketisation, receives scant attention.

Policy makers claim that an efficient, productive economy raises living standards, yet in making this claim, there is scarce reflection upon what living standards are. By connecting the goal of increasing living standards to economic performance, which itself is enhanced through marketisation, policy makers construct living standards as an economic concept, measurable through economic data. In this manner, living standards are constructed as achievable in markets, by firms. According to this discourse, firms compete for consumer preferences and are compelled by competitive pressures to reduce the prices of their goods and services, to increase product quality and diversity, and to innovate. Therefore, markets are claimed to raise living standards, as consumers are able to choose between high quality, affordable products and services which meet their needs and, consequently, raise their standard of living. Liberalising markets, and creating a business-friendly environment, is therefore constructed as an essential aspect of the process of raising living standards. According to this construction, living standards are raised by firms, acting efficiently, in competitive markets. This narrative is dependent upon the assumptions that firms lack power in markets, that firms will pass on any efficiency gains to consumers, and that consumer choice is a genuine form of power which compels firms to act in consumers' best interests.

However, this narrative is also built upon the construction of the consumer: for this narrative to hold, people must view their living standards in terms of their experience as consumers and in relation to the good and services that they consume. Consumption becomes the mechanism whereby people can enhance their standard of living. Portraying citizens as consumers enables living standards to be constructed as materialistic, and measurable through knowledge produced and communicated by markets. Pusey critiqued this focus on markets as a means of enhancing consumers' living standards, by citing the welfare economics concept of living standards as reflecting aggregate utility "measured in dollars by the market" (Pusey 1996, p. 72). In this sense, markets create and measure living standards according to the knowledge which markets understand, such as consumer choices and prices (Pusey 1996, p. 74). Thus, the notion of living standards, and the knowledge with which it is constructed and assessed, is not interchangeable with other concepts such as quality of life and well-being, as some policy documents suggest (NCOA 2014). Pusey (1996, p. 72) highlights the dissimilarities between "living standards" and "quality of life", arguing that life priorities of people are only indirectly related to income or money, or consumables (Pusey 1996, 2003a, b). This illustrates the disconnection between policy conceptualising living standards in terms of utility and what it is that people actually desire as part of a meaningful, fulfilling life. Yet policy discourses do not highlight the problems associated with this disjuncture. Policy makers construct living standards as a vague, yet desirable goal which can be achieved through greater efficiencies, as firms compete to satisfy consumer needs. This turns attention away from questions about what efficiency is, or what conception of "living standards" should be enhanced, towards the narrowed question of what policy change is required to maximise efficiency.

Marketising Society

This chapter shows how Australian mainstream policy discourses have framed marketisation of social and economic life previously controlled by the state as being beneficial for society. The rationality of efficiency, competition and markets as being the central truths around which society should be organised, is integral to this shift in discourse. In effect, this discursive shift has portrayed the relationship between the markets, firms, and consumers as a symbiotic one. The needs of firms, such as liberalised markets and reduced regulation, have come to be accepted as the needs of consumers. This construct reduces society to consisting of "businesses and consumers" (DPMC 2014, p. vi). Therefore, policy discourses claim that society will benefit from increased marketisation, as this compels greater resource efficiency, as firms, governed by consumers and competitive markets, will seek to maximise resource use. In turn, this raises productivity, on the one hand, and consumer satisfaction, on the other. As a result, economic performance improves, as measured in statistical data such as GDP and multifactor productivity, raising living standards. However, the conception of living standards relates to people's experiences as consumers, rather than their experiences as part of a functioning, thriving society. In addition, the question of how these gains are distributed is rarely addressed in mainstream policy discourses. It is claimed that marketisation creates opportunities for those with the self-reliance, ingenuity, and adaptability to take them; however, this construct individualises people, who are cast as being responsive to incentives, rather than the collective interests of society. In this sense, society, and particularly the state's role in supporting society, is increasingly marginalised, in favour of the individual. The role of the individual is conceptualised in terms of their experiences as a consumer. Effectively, society itself is structured out of this picture, as an absent actor.

Consumers are integral to marketisation, as consumers communicate their preferences to firms, which directs how firms should allocate their resources. Thus, as a governmental technology, the consumer is central to making efficient markets. People's roles as workers and producers are cast as a secondary concern. Rather than protecting jobs or supporting workers to retain job security and the right to meaningful work, policy discourses portray the role of the state as being focused on nurturing the competitive, business-friendly environment, which is presumed to lift efficiency and economic performance. Thus, the process of producing is cast as having little relevance, provided that markets are able to determine the most efficient, most productive process for converting resources to products to be sold in markets. In this regard, workers' rights, or farmers' rights to remain on the land and to receive adequate remuneration for their commodities, is constructed as being relatively insignificant in regard to the greater purpose of maximising efficiency. Regulation protecting the interests of cohorts such as workers or farmers is framed as a barrier to efficiency. In this context, regulation harms society, which is considered in terms of consumers and businesses. By virtue of this, the interests of groups such as workers or farmers are portrayed as being counter to the national interest, which is to maximise efficient resource use.

On the other hand, firms are constructed as being central to this overarching aim. Marketisation of industries such as farming, human services, or social services is portrayed as being essential for Australia's economic performance. This shift facilitates firms' entry to these industries, whose superior efficiency and responsiveness to consumers will contribute to the improved productivity and efficiency of resource use, and ultimately enhance Australia's economic performance. This construction has externalised sections of Australian society involved in the production process, such as workers and farmers, and has contributed to widening economic inequalities within Australia.

References

Abbott, T., (2013), *Media Release: Review of Competition Policy*, December 4, 2013, [online] http://bfb.ministers.treasury.gov.au/media-release/014-2013/ [Accessed July 14 2015].

Aulich, C., and O'Flynn, J., (2007), From public to private: The Australian experience of privatisation, *Asia Pacific Journal of Public Administration*, 29(2), pp. 163–171.

Baird, M., Cooper, R., and Ellem, B., (2009), Low-paid women: The impact of regulatory change in Australia, *Industrial Relations Journal*, 40(5), pp. 393–407.

Baker, C., (2018), The nation building state retreats: An Australian case study in the changing role of the state, *Journal of Rural Studies*, 62, pp. 146–155.

Baumol, W., and Ordover, J., (1985), Use of antitrust to subvert competition, Journal of Law and Economics, 28(2), *Antitrust and Economic Efficiency: A conference sponsored by the Hoover Institution* (May 1985), pp. 247–265.

Beer, A., Bentley, R., Baker, E., Mason, K., Mallett, S., Kavanagh, A., and LaMontagne, T., (2016), Neoliberalism, economic restructuring and policy change: Precarious housing and precarious employment in Australia, *Urban Studies*, 53(8), pp. 1542–1558.

Benach, J., Vives, A., Amable, M., Vanroelen, C., Tarafa, G., and Muntaner, C., (2014), Precarious employment: Understanding an emerging social determinant of health, *Annual Review of Public Health*, 35, pp. 229–553.

Bork, R., (1978), *The antitrust paradox: A policy at war with itself*, New York: Free Press.

Bork, R., (1967), The goals of antitrust policy, *The American Economic Review*, 57(2), Papers and Proceedings of the Seventy-Ninth Annual Meeting of the American Economic Association (May 1967), pp. 242–253.

Campbell, I., (2008), *Australia: Institutional changes and workplace fragmentation*, in Lee, S., and Eyraud, F., (ed.) Globalisation, Flexibilisation and Working Conditions in Asia and the Pacific, pp. 115–152, London: Woodhead Publishing Limited.

Campbell, K., (1974), Rural Policy in Australia: 1974 style, *The Australian Journal of Agricultural Economics*, 18(3), pp. 157–170.

Colebatch, H., (2002), Government and governmentality: Using multiple approaches to the analysis of government, *Australian Journal of Political Science*, 37(3), pp. 417–435.

Coleman, W., and Skogstad, G., (1995), Neo-liberalism, policy networks and policy change: Agricultural policy reform in Australia and Canada, *Australian Journal of Political Science*, 30(2), pp. 242–263.

Cooper, R., and Ellem, B., (2008), The neoliberal state, trade unions and collective bargaining in Australia, *British Journal of Industrial Relations*, 46(3), pp. 532–554.

Demsetz, H., (1973), Industry structure, market rivalry, and public policy, *The Journal of Law and Economics*, 16(1), pp. 1–9.

Department of the Prime Minister and Cabinet, (2014), *Industry innovation and competitiveness agenda: An action plan for a stronger Australia*, [online] https://www.dpmc.gov.au/sites/default/files/publications/industry_innovation_competitiveness_agenda.pdf [Accessed October 5 2016].

Derkatch, C., and Spoel, P., (2017), Public health promotion of "local food": Constituting the self-governing citizen-consumer, *Health*, 21(2), pp. 154–170.

Fairbrother, P., Svenson, S., and Teicher, J., (1997), The withering away of the Australian State: Privatisation and its implications for labour, *Labour & Industry: A Journal of the Social and Economic Implications of Work*, 8(2), pp. 1–29.

Gerritsen, R., (1987), Why the 'uncertainty'? Labor's failure to manage the 'rural crisis', *Politics*, 22(1), pp. 47–59.

Gray, E., Oss-Emer, M., and Sheng, Y., (2014), *Australian agricultural productivity growth: Past reforms and future opportunities*, Australian Bureau of Agricultural and Resource Economics and Sciences, Research Report 14.2, Canberra, February.

Griggs, P., (2002), Changing rural spaces: Deregulation and the decline of tobacco farming in the Mareeba-Dimbulah Irrigation Area, Far North Queensland, *Australian Geographer*, 33(1), pp. 43–61.

Guthman, J., and Brown, S., (2016), I will never eat another strawberry again: The biopolitics of consumer-citizenship in the fight against methyl iodide in California, *Agriculture and Human Values*, 33, pp. 575–585.

Harper, I., (2015), *Speech to CEDA*, [online] http://competitionpolicyreview.gov.au/2015/04/14/speech20150414/ [Accessed July 14 2015].

Harper, I., Anderson, P., McCluskey, S., and O'Bryan, M., (2015), *Competition Policy Review: Final report*, [online] http://competitionpolicyreview.gov.au/final-report/ [Accessed July 30 2015].

Harris, S., (1974), *The principles of rural policy in Australia: A discussion paper*, Canberra: Australian Government Publishing Service.

Healy, J., (2016), The Australian labour market in 2015, *Journal of Industrial Relations*, 58(3), pp. 308–323.

Higgins, V., and Lockie, S., (2002), Re-discovering the social: neo-liberalism and hybrid practices of governing in rural natural resource management, *Journal of Rural Studies*, 18, pp. 419–428.

Hilmer, F., Rayner, M., and Taperell, G., (1993), *National Competition Policy Review*, [online] http://ncp.ncc.gov.au/docs/National%20Competition%20Policy%20Review%20report,%20The%20Hilmer%20Report,%20August%201993.pdf [Accessed July 30 2015].

Howard, J., (1995a), *Fair Australia: John Howard Headland Speech*, presented to the Australian Council of Social Service, October 13, 1995, [online] http://australianpolitics.com/1995/10/13/fair-australia-howard-headland-speech.html [Accessed July 30 2015].

Howard, J., (1995b), *The role of government*, Speech presented at The Menzies Research Centre, June 6 1995, [online] http://australianpolitics.com/1995/06/06/john-howard-headland-speech-role-of-govt.html [Accessed August 29 2016].

Humphrys, L., (2018), Simultaneously deepening corporatism and advancing neoliberalism: Australia under the Accord, *Journal of Sociology*, 54(1), pp. 49–63.

Industry Assistance Commission, (1988), *The Wheat Industry*, Report No. 411, AGPS, Canberra.

Industry Commission, (1991), *Statutory marketing arrangements for primary products*, Report No. 10, Canberra: Australian Government Publishing Service.

Jobs for New South Wales, (2016), *Jobs for the future: Adding 1 million rewarding jobs for NSW by 2036*, [online] http://www.jobsfornsw.com.au/__data/assets/pdf_file/0020/90740/Jobs-for-the-future-fullreport-August-2016.pdf [Accessed May 30 2017].

Jose, J., and Burgess, J., (2005), Working Nation: Context and consequences, *Journal of Economic and Social Policy*, 9(2), pp. 1–14.

Keating, P., (1992), *One nation: Statement by the Prime Minister*, Canberra: Australian Government Publishing Service.

Kremers, R., and Brassett, J., (2017), Mobile payments, social money: Everyday politics of the consumer subject, *New Political Economy*, 22(6), pp. 645–660.

Larner, W., (2006), *Neoliberalism: Policy, ideology and governmentality*, in de Goede, M. (ed.), International political economy and poststructural politics, pp. 199–218, United Kingdom: Palgrave Macmillan.

Le Heron, R., and Roche, M., (1999), Rapid regulation, agricultural restructuring, and the reimaging of agriculture in New Zealand, *Rural Sociology*, 64(2), pp. 205–218.

Magnan, A., (2015), The financialisation of agri-food in Canada and Australia: Corporate farmland and farm ownership in the grains and oilseed sector, *Journal of Rural Studies*, 41, pp. 1–12.

Martin, W., (1990), Public Choice Theory and Australian Agricultural Policy Reform, *The Australian Journal of Agricultural economics*, 34(3), pp. 189–211.

Mauldon, R., and Schapper, H., (1974), *Australian farmers under stress in prosperity and recession*, Perth: University of Western Australia Press.

McDonald, C., and Marston, G., (2005), Workfare as welfare: Governing unemployment in the advanced liberal state, *Critical Social Policy*, 25(3), pp. 374–401.

McDonald, C., and Marston, G., (2001), *Fixing the niche? Rhetorics of the community sector in the neo-liberal welfare regime*, Paper presented to the workshop, 'Social entrepreneurship: whose responsibility is it anyway?, November 21 2001, University of Newcastle.

McKinsey Australia, (2014), *Compete to prosper: Improving Australia's global competitiveness*, [online] http://www.mckinsey.com/~/media/mckinsey%20offices/australia/pdfs/compete_to_prosper.ashx. [Accessed August 10 2016].

Meagher, G., and Wilson, S., (2015), The politics of market encroachment: Policymaker rationales and voter responses, in Meagher, G., and Goodwin, S., (eds.), *Markets, rights and power in Australian social policy*, pp. 29–96, Sydney: Sydney University Press.

Morley, C., and Ablett, P., (2016), A critical social work response to wealth and income inequality, *Social Alternatives*, 35(4), pp. 20–26.

Morris, A., and Wilson, S., (2014), Struggling of the Newstart unemployment benefit in Australia: The experience of a neoliberal form of employment assistance, *The Economic and Labour Relations Review*, 25(2), pp. 202–221.

Murdoch, R., (2013), *Free market is a fair market: Murdoch*, Speech presented to the Institute of Public Affairs, April 4th 2013, The Australian, [online] http://www.theaustralian.com.au/business/media/push-to-regulate-press-ill-conceived-murdoch/news-story/70177400a9b3aaf859a45bd96049873d [Accessed October 31 2017].

National Commission of Audit, (2014), *Towards responsible government: The report of the National Commission of Audit*, Phase One, Commonwealth of Australia, Canberra, [online] http://www.ncoa.gov.au/REPORT/docs/phase_one_report.pdf [Accessed October 5 2016].

National Competition Council, (2009), *Submission to Productivity Commission Inquiry into Wheat Export Marketing Arrangements*, [online] http://ncc.gov.au/images/uploads/SubPCWEM-001.pdf [Accessed October 26 2016].

National Farmers Federation, (2013), *Red tape in Australian agriculture: NFF Issues Paper*, [online] file://ntapprdfs01n01.rmit.internal/el4/E68334/NFF%20Issues%20Paper%20-%20Red%20Tape%20in%20Australian%20Agriculture%20Sept2013.pdf [Accessed December 20 2016].

O'Keeffe, P., and Neave, M., (2017), Experiences of wheat growers in Victoria's western Wimmera following deregulation of the Australian wheat export market, *Rural Society*, 26(1), pp. 1–18.

Productivity Commission, (2016), *Digital disruption: What do governments need to do?*, Commission Research Paper, Canberra.

Productivity Commission, (2013), *Productivity update*, May 2013, [online] file://ntapprdfs01n01.rmit.internal/el4/E68334/2013-pc-productivity-update.pdf [Accessed May 30 2013].

Productivity Commission, (2005a), *Review of National Competition Policy reforms: Productivity Commission Inquiry Report*, No. 33, 28 February 2005, Commonwealth of Australia, Canberra.

Productivity Commission, (2001), *Structural Adjustment – Key policy issues, Commission Research Paper*, AusInfo, Canberra.

Powell, D., (2018), Governing the (un)healthy child-consumer in the age of the childhood obesity crisis, *Sport, Education and society*, 23(4), pp. 297–310.

Prince, R., Kearns, R., and Craig, D., (2006), Governmentality, discourse and space in the New Zealand health care system, 1991–2003, *Health and Place*, 12, pp. 253–266.

Pritchard, B., (2005), Implementing and maintaining neoliberal agriculture in Australia, *International Journal of Sociology of Agriculture and Food*, 13(2), pp. 1–14.

Pusey, M., (2016), *Introduction to the Roundtable '25 Years since Michael Pusey's Economic Rationalism in Canberra'*, 2016 conference of The Australian Sociological Association, Australian Catholic University, Melbourne, Australia.
Pusey, M., (2003a), *An Australian story: The troubling experience of economic reform*, [online] https://www.aph.gov.au/binaries/senate/pubs/pops/pop40/pusey.pdf [Accessed February 23rd 2017].
Pusey, M., (2003b), *The experience of middle Australia*, Port Melbourne, Australia: Cambridge University Press.
Pusey, M., (1998a), Economic rationalism, human rights and civil society, *Australian Journal of Human Rights*, 4(2), pp. 131–152.
Pusey, M., (1998b), Women & Families: The impact of economic restructuring on women and families: Preliminary findings from the middle Australia project, *Australian Quarterly*, 70(4), pp. 18–27.
Pusey, M., (1996), Economic rationalism and the contest for civil society, *Thesis Eleven*, 44, pp. 69–86.
Pusey, M., (1991), *Economic Rationalism in Canberra: A nation building State changes its mind*, Melbourne: Cambridge University Press.
Quiggin, J., (2009), Six refuted doctrines, *Economic Papers*, 28(3), pp. 239–248.
Quiggin, J., (2002), Economic governance and microeconomic reform, in Bell, S., (ed.), *Economic Governance and Institutional Dynamics*, (pp. 158–177), South Melbourne: Oxford University Press.
Quiggin, J., (1995), Does privatisation pay?, *Australian Economic Review*, Second Quarter, pp. 23–42.
Ryan, M., (2015), Contesting 'actually existing' neoliberalism, *Journal of Australian Political Economy*, 76, pp. 76–102.
Ryan, N., (2005), A decade of social policy under John Howard: Social policy in Australia, *Policy and Politics*, 33(3), pp. 451–460.
Scourfield, P., (2007), Social care and the modern citizen: Client, consumer, service user, manager and entrepreneur, *British Journal of Social Work*, 37, pp. 107–122.
Spies-Butcher, B., (2014), Marketisation and the dual welfare state: Neoliberalism and inequality in Australia, *The Economic and Labour Relations Review*, 25(2), pp. 185–201.
Stanford, S., and Taylor, S., (2013), Welfare dependence or enforced deprivation? A critical examination of White neoliberal welfare and risk, *Australian Social Work*, 66(4), pp. 476–494.
Stilwell, F., and Jordan, K., (2007), Economic inequality and (Un)happiness, *Social Alternatives*, 26(4), pp. 16–21.
Stilwell, F., (1994), Working nation: From green to white paper, *Journal of Australian Political Economy*, 33, pp. 111–123.
Stilwell, F., (1991), Wages policy and the Accord, *Journal of Australian Political Economy*, 28, pp. 27–53.

Turnbull, M., (2015), *Transcript: Interview with Neil Mitchell on 3AW*, October 6 2015, [online] https://www.pm.gov.au/media/2015-10-06/interview-neil-mitchell-3aw [Accessed August 29 2016].

Villadsen, K., and Wahlberg, A., (2015), The government of life: Managing populations, health and scarcity, *Economy and Society*, 44(1), pp. 1–17.

Watts, R., (2016), *'Running on empty': Australia's neoliberal social security system, 1988–2015*, in Mays, J., Marston, G., and Tomlinson, J., Basic income in Australia and New Zealand: Perspectives from the neoliberal frontier, pp. 69–91, Hampshire: Palgrave Macmillan.

Western, M., Baxter, J., Pakulski, J., Tranter, B., Western, J., van Egmond, M., Chesters, J., Hosking, A., O'Flaherty, M., and van Gellecum, Y., (2007), Neoliberalism, inequality and politics: The changing face of Australia, *Australian Journal of Social Issues*, 42(3), pp. 401–418.

Wilson, S., Spies-Butcher, B., Stebbing, A., and St John, S., (2013), Wage-earners' welfare after economic reform: Refurbishing, retrenching and hollowing out of social protection in Australia and New Zealand, *Social Policy and Administration*, 47(6), pp. 623–646.

Wilson, S., and Turnbull, N., (2005), Wedge politics and welfare reform in Australia, *Australian Journal of Politics and History*, 47(3), pp. 384–402.

CHAPTER 4

Productivism, Financialisation, and the "Good Farmer": Constructing a Rational, Governable Farming Sector

Introduction

This chapter addresses the concept of productivity in the context of structural adjustment policies implemented by the Australian government in recent decades. As I argue in this chapter, this focus reduces the value of farming, and of farmers, to the capacity to be productive. As in the normalisation of competition as a process of creating winners and losers, I show that structural adjustment normalises farmer exits as an essential step in maximising industry efficiency and productivity. I then develop this idea further by arguing that these reductionist policy constructions externalise the social and environmental consequences of the neoliberalisation of Australian agricultural and rural industries. Drawing on policy discourses around agriculture, specifically relating to structural adjustment and farmers' identity, I explore the construction of farming as a reducible, calculable function which the state is able to act upon to meet its objective of maximising productive resource use. In this manner, governance in agriculture is made possible through the actions, behaviours, and values of individualised farmers.

Policy makers' focus on allocative efficiency underpins this shift. This aspect of economic efficiency is applied to rural and agricultural policy to reallocate resources to the most efficient resource managers. Policy makers claim this is necessary to facilitate the exits of the least efficient actors from the industry, thus liberating resources controlled by these farmers, such as farm land and machinery. Consequently, it is argued the most efficient

P. O'Keeffe, *Making Markets in Australian Agriculture*,
https://doi.org/10.1007/978-981-13-3519-8_4

farmers will acquire these resources, expand their farming operation, and put these resources to more productive use. This is portrayed as beneficial for Australian society, as it improves industry efficiency and productivity. This conceptualises farming as amounting to a collection of resources, from which the only question is how to maximise the productive use of these resources.

As I show in this chapter, this shift is made possible, in part, by the reconstruction of farming and farmer identity. The value of farming is reduced to its capacity to convert resources into commodities. Questions around agricultural organisation become focused on enhancing the efficiency of this process. By maximising the efficiency of this process, farmers enhance their contribution to Australian society. This helps make structural adjustment possible. Farmers that are least able to do this efficiently are constructed as not only poorly skilled farmers but also an impairment on societal welfare. The exit of these farmers from the industry is therefore rationalised as a positive shift. This construction overlooks the valuable role of farmers in supporting rural communities and caring for the land, and also the relationship between farmers, their land, and their histories in farming.

Moreover, using Dean's (1999) conceptualisation of technologies of agency, I build on the work of Higgins et al. (2015), Higgins (2001, 2002), Lockie and Higgins (2007), Higgins and Lockie (2002), and Argent (2005), to draw upon the construction of the good farmer as a self-reliant, rational, and business-minded actor. I show how the shaping of this identity has been used to facilitate policy shifts such as the structural adjustment packages, the deregulation of the wheat export market, and the enhanced financialisation of farming in Australia. Policy makers have utilised this construct of the good farmer, to frame these policy shifts as being essential, and in farmers' best interests. Good farmers, it is claimed, are not reliant upon state support, desire the freedom and choice of liberated markets, and, critically, have the entrepreneurial focus to reorganise their farm business to make their land attractive to private investors. This presents the good farmer as one who is willing to embrace the business-friendly environment of deregulated markets and increased private ownership of farmland. As I suggest, this facilitates the development of a policy environment which further shifts control over agriculture and farmland to private interests such as agri-business firms and institutional investors.

Productivism and the National Interest

Productivism is deeply ingrained within Australian agricultural policy making, as an essential approach to managing agricultural land. In 1971, the CEDA published a collection of works by agricultural economists entitled "Productivity". Critically, in his contribution "The Government, the economy, and productivity", Nevile (1971, p. 3) claimed that "government policies must be such as will lead to productive resources being used in those ways which are the most efficient from the point of view of the economy as a whole." This is an important statement in shifting the purpose and value of the farm sector. If farmers enter this picture, it is as resource managers. This approach contends that agricultural policy should view farms as a collection of resources, from which the challenge is to extract maximum productivity. Farmers are then defined by their capacity to be efficient, and to contribute to this broader ambition. This reconstruction of the farmer is critically important to the overall picture of agricultural deregulation, and has significant implications for Australian farmers and their communities. Furthermore, this raises the concept of the productivity of agricultural industries, and how farmers' contribution to the national economy is measured and assessed. Since the early 1970s, policy makers have sought to facilitate the organisation of the farm sector to maximise productivity, through policies such as the Rural Reconstruction Scheme, the Structural Adjustment packages and Rural Adjustment Schemes, the 1992 National Drought Plan and the 1997 Agriculture—Advancing Australia package, and the more recent 2015 Agricultural Competitiveness White Paper.

Agricultural productivity is described by Nossal and Gooday (2009, p. 4) as reflecting "changes in the efficiency of converting inputs into outputs", and is pursued as a policy objective under the assumption that this will lead to increased living standards (Nossal and Gooday 2009). Productivism, as an approach to farming which aims to maximise productivity growth, is argued to result from the expanded use of technology, from increases in farm size and scale, and through the reallocation of resources to more efficient farm managers, which has been facilitated through structural adjustment programmes implemented by the Australian government in the past five decades (Nossal et al. 2009; Knopke, O'Donnell, and Shepherd 2000; RIRDC 2007; Gray et al. 2014; Productivity Commission 2005; Lawrence et al. 2013; Dibden, Potter, and Cocklin 2009). Structural adjustment as a productivity driver is associated with the concept of allocative efficiency, outlined by Hilmer, Rayner, and Tapparell (1993, p. 4) as being maximised when "resources used to produce a set of goods or services are allocated to their highest valued uses." In relation to agriculture, allocative efficiency is increased when resources are managed by the most efficient, most productive farm managers (Gray et al. 2014). Resultantly, it is argued that productivity increases with the exit of least efficient, least productive farmers from the industry (Nossal and Gooday 2009; Gray et al. 2014). Fundamentally, this approach to policy is made possible by constructing farmers as resource managers, by constructing farms as "the community's resources", and by focusing policy upon the maximisation of the productive use of farm resources. As I show in the next section, this approach separates farmers from their social contexts.

Rationalising Farmers' Value as Productive Units

In recent decades, Australian agricultural policy has shifted from protecting farmers, towards protecting markets. This change reimagines governments' role in agriculture, frequently explained through policy discourses as creating the environment for markets to function. This, it is claimed, enhances efficiency, by connecting farmers with markets. As Larner (2000, p. 245) explains, this conceptualisation of the state's role is based upon the assumption that governments should create an environment "where active citizens will exercise responsibility for themselves and their families". Accordingly, policy discourses construct freedom as the capacity of the individual to make responsible, informed choices, unimpeded by the state (Lockie and Higgins 2007). This construction is central to arguments which have contributed to the deregulation of wheat export marketing and other marketing arrangements underpinning dairy, wool, barley, and egg industries (Gill 2011; Productivity Commission 2005; Industry Commission 1991). Collectivist approaches to agricultural organisation, which had underpinned Australian agricultural policy, were problematised by policy discourses as a restraint on freedom of the individual. In addition, collectivism was portrayed as a policy approach which rewarded mediocre performance and reduced the positive impact of incentives. This construction was at odds with policies such as statutory wheat marketing emphasised the collective interests of farmers. Policy makers argued the redistributive function of statutory marketing arrangements prevented farmers from maximising their returns, while restricting freedom and choice. As such, it was claimed that state intervention in agricultural marketing restricted market and industry efficiency. Discursive shifts which constructed farmers as individuals, rather than as a collective, were therefore essential in reshaping the state's responsibility as creating an environment which enabled self-interested farmers to maximise their returns, through exercising freedom and choice.

Policy makers pursue structural adjustment, believing this facilitates the reallocation of resources to the most productive farm managers, enhancing productivity, competitiveness, resilience, and self-reliance of the farm sector. This is reflected by Gray et al. (2014, p. 5), who describe the role of structural adjustment in increasing allocative efficiency:

> At an industry level, ongoing resource reallocation in an important source of productivity gains...In particular, exits of less efficient farm businesses release scarce resources for use by more efficient farms, which are able to expand and increase productivity, increasing the efficiency of the resource as a whole.

The distant terminology which pervades this discourse seeks to render agriculture as eminently calculable (Miller and Rose 1990, p. 5). Farms are described as farm units, or as scarce resources, from which policy must aim to extract maximum productivity. The policy question becomes centred on how to achieve greatest productivity from the national resource base, with farmers operationalised as resource managers. This mechanistic approach necessarily views farmers as atomistic individuals whose value to society is considered in terms of their capacity to contribute to the productive management of the resources they control. The construction of farmers as resource managers decontextualises farmers from their communities and land. Thus, policy discourses construct the exits of the least efficient resource managers from the industry as having no perceptible negative impacts. Whereas policy discourses claim this process strengthens communities and builds resilience, I argue these reductionist conceptions of community, strength, and resilience are narrow, and fail to capture the contribution of farmers to their communities. Subjectivity is rendered incalculable, and therefore unknowable. The more subjective farming values, such as family history in farm properties and the community, for example, are treated as external to the primary ambition of maximising allocative efficiency.

Farmers are reconstituted as tools assisting the productive use of resources. In this process, the value of farming first needed to be narrowed to questions that could be answered by quantitatively measuring efficiency and productivity. In this policy context, the value ascribed to the productivity of resource use significantly overshadows the broader impact of this shift upon farmers and their communities. Policy discourses construct farmers' role in the industry as replaceable, while according minimal significance to their community function, beyond the transactions associated with their business. This is framed as a rational response to the realities of compressed terms of trade and over-supply of commodities. In that regard, while the individualisation of farmers facilitates structural adjustment, this is made possible by the rationalist reconstruction of farmers and farming as calculable and administrable. Thus, the role of government shifts to focus on maximising productive use of farm resources, through individualised, self-reliant farmers.

The reallocation of resources articulated in the structural adjustment policies of the 1980s is made possible through measuring farmers against their capacity to manage resources efficiently. Policy discourses suggest that the more farmers are able to maximise the efficiency of this process, the more value they add to Australian society. The farmers who are least able to meet this aim are constructed as inefficient, poorly skilled, and as barriers to industry efficiency and consumer well-being. Thus, farmer exits are rationalised as a positive shift.

In turn, government intervention is framed as unjustifiably supporting substandard farmers. Smart farmers are cast as those which adopt technology, avoid government support, and favour market deregulation. This reflects the construction of good farmers as active agents, who adopt the language of self-reliance and responsibility, and develop their capacity to enhance productivity and stay viable. Conversely, the notion of the "grossly inefficient family farm" contributes to the perception that deregulation is necessary and overdue (Financial Review 1992). Policy discourses construct farm unprofitability as an individual issue, stemming from poor farm management, outdated practices, lack of adaptability, and incapacity or unwillingness to learn. The structural causes of this problem are not addressed.

Rather, policy makers view the exit of unprofitable farmers from farming as an essential, and inevitable, process of resource reallocation. The construct of the unprofitable farmer is therefore necessary to make the process of structural adjustment appear as a common-sense solution to problems of industry inefficiency. Their inefficiency is portrayed as a personal failing. Their continued existence in farming is cast as an impediment to the governments' aim of maximising productive resource use. The role of government is repositioned to focus on transitioning these farmers out of the industry, causing their resources to be reallocated to more productive and efficient actors.

Market Liberalisation and "the Good Farmer"

The shift in values underpinning wheat industry policy has contributed to wheat industry deregulation. Equally important, policy discourses have reshaped farmer identity to accord with these values (Cheshire and Lawrence 2005; Higgins and Lockie 2002). The construction of the "good farmer" is central to this, which, I suggest, has been operationalised to help facilitate wheat export market deregulation. The good farmer is considered in terms of their self-conceptualisation as a rational economic actor. The good farmer views themselves as a business person, assesses the costs and benefits associated with their business, and sensibly makes decisions on this basis. This farmer employs common sense, and according to this discourse, common-sense farming is that which maximises returns. Assessments of the wheat industry, and value of regulation in particular, aim to communicate to growers how much they could benefit financially from a policy shift which reduces regulation (Irving et al. 2000; Allen Consulting 2000; Royal Commission into Grain Transport, Handling and Storage 1988). Policy discourses rationalise and calculate the costs and benefits of regulation. This approach portrays quantifiable information as concrete, credible, and value-free evidence. Policy discourses frame good policy, and good business decisions, as based upon assessment of this evidence. As this evidence is constructed to show that deregulation will deliver higher returns to growers, and growers are framed as rationally concerned with maximising their returns, good farmers are constructed as those which want to "getting on with business" and support deregulation of the wheat export market (Productivity Commission 2010, p. 84).

Good farmers are portrayed as competitive, in the sense that they relish competition with other farmers, and fiercely individualistic. Yet statutory wheat marketing considers growers as a collective, precluding competition between farmers and essentially delivering equal returns to growers. Mainstream economic and competition policy discourses through the 1980s and 1990s frequently lament the injustice of such arrangements. Good farmers, it is claimed, are unjustly required to support the less-efficient, less-skilled farmers through statutory marketing, as costs and returns are redistributed among growers, regardless of how efficient or how skilled they may be. This is argued to reduce incentives, and moreover, to place an unfair burden on the most efficient farmers. However, this also prevents growers from utilising their skills and knowledge to maximise their wheat prices. Thus, a good wheat farmer is framed as one that wants to develop these skills and will be prepared to accept their loss should they fail to build these skills and use them effectively. This is portrayed as fair: the market determines, without prejudice, which farmers succeed or fail based on their skills, knowledge, organisation, and judgement.

This construction assumes that all will have equal ability and resources to market their wheat successfully. In addition, some growers may experience the freedom of liberalised markets as a burden. However, this is not necessarily a concern for policy makers, who want to encourage growers who are prepared to accept responsibility for their wheat marketing. Those farmers who are incapable of developing the necessary skills and attitudes to market their own wheat successfully are constructed as poor farm managers, who do not deserve ongoing government assistance to remain in farming. Their exits from farming are portrayed as an essential step in maximising productive resource use. Good farmers remaining in the industry subsume the resources that are made available through these exits and put them to more productive use. Thus, the good farmers, for whom wheat export market deregulation was intended, operationalise the state's central aim of increasing efficient use of the community's productive resources.

Importantly, this creation of the good farmer, as a rational, self-reliant actor intent on maximising their returns, decontextualises farmers from their connections to their land and communities. This creates an "abstracted and aspatialised" conception of farming (Pritchard 2005, p. 110), which externalises the social implications of wheat export market deregulation in particular and centres policy restructuring on the impact upon individual farmers, and their capacity to maximise their returns by exercising choices in liberalised markets.

Agricultural Competitiveness White Paper: Creating Dependent Farmers?

Farming has become increasingly rationalised, with the social impacts of deregulation erased from the consciousness of policy makers. Thus, efficiency, productivity, and profitability have been constructed as the sole sources of legitimacy for farmers. The separation of farmers from their communities, through rationalist discourses, reinforces this construction. Governments, perceiving their role to be as market stewards, focus upon key economic indicators to gauge the contribution of the farm sector to the national economy. Supporting farmers to remain in the industry, and to remain in their communities, is framed as a barrier to efficient market operation. Thus, preserving the purity of the economic environment is "what matters". Farmers, if they want to keep farming, must adapt to this environment, as suggested by then Prime Minister Tony Abbott and Minister for Agriculture, Barnaby Joyce, in the introduction to the 2015 Agricultural Competitiveness White Paper (2015, p. ii):

> Australia cannot be complacent. Consumer demand shifts and we are not the only country competing for lucrative overseas markets. The agriculture sector, like all business, must keep up and commit to continuous improvement. But our farmers have the ingenuity and enterprise to do this, as they always have.

Good farmers, therefore, are those who can respond to these challenges. Farmers' failure to adapt is not perceived as sign that there is a problem with this competitive environment. Rather, adjustment is received by policy makers as a sign that this business-friendly market environment is working efficiently. Adjustment is constructed as the necessary exits of farmers who are unable to "keep up" with the pace of change. Thus, farmers' survival in this environment is entirely their responsibility.

The Agricultural Competitiveness White Paper (White Paper) was announced by the Australian government as a "a clear strategy to improve the competitiveness and profitability of the agriculture sector, boosting its contribution to trade and economic growth, and building capacity to drive greater productivity through innovation" (Commonwealth of Australia 2013, p. 1). As the White Paper argues, increasing the competitiveness of the sector is essential, as "stronger farmers mean a stronger Australian economy" (Commonwealth of Australia 2015, p. 1). This statement conceptualises strength in terms of farm profitability, implying that farmers experiencing financial difficulty lack strength and resilience. This emphasises the White Paper's focus on the individual and its lack of reflection on the structural causes of farm profitability. For example, the White Paper acknowledges the consolidation of commodities markets, which is particularly evident in the deregulated wheat export market

(Commonwealth of Australia 2015, p. 23). However, rather than critiquing this outcome of deregulation, the White Paper instead focuses on enabling farmers to succeed in this environment, principally, through attracting investment from the private sector. Despite assurances that wheat export market deregulation would empower growers through choice and freedom, this document implies that the good farmer will overlook the apparent failures of this policy shift and, instead, reconstruct their farm business to attract private investment.

The White Paper constructs farms as investment targets. Attracting private investment is framed as an essential part of "modern" farming, with farmers' ability to do this central to their role as farmers (Commonwealth of Australia 2015, p. 7). In turn, the government considers its role to be centred on providing farmers with the skills to attract investment. Farmers, it is argued, need to attract "external non-bank capital", through enhancing their capacity to "demonstrate value and provide investable products that allow external investment" (Commonwealth of Australia 2015, p. 47). The White Paper explains that domestic superannuation funds' reluctance to invest in farming is due to the long-term nature of farm investments, raising the question of whether the White Paper is suggesting that farmers should revise their farming models to deliver shorter-term gains for external investors.

This focus on private investment and construction of farms as investment targets fundamentally changes the relationship between farmers and their farms. Farms are portrayed as investments, and the farmers' role is portrayed as shifting from maximising the productive use of their farm resources, towards attracting capital investment from external sources. In this sense, the governments' aim is to encourage greater private sector investment into agriculture and, in the wake of the end of Australia's mining boom, stimulate investment into the Australian economy. However, this appears to be endorsing the financialisation of farm land, without sufficiently considering the implications of this shift.

Financialisation and the Loss of Control

Financialisation of global commodities markets, farm land, and agri-food industries has emerged in recent years, as financial actors such as private equity groups and hedge funds become prominent investors in agriculture (Sippel, Larder, and Lawrence 2017; Salerno 2017; Clapp 2015; Martin and Clapp 2015; Fairbairn 2014; Burch, Dixon, and Lawrence 2013).

Profit making opportunities in agriculture are increasingly occurring through financial mechanisms, rather than through the trade in tangible commodities (Sippel, Larder, and Lawrence 2017; Murphy et al. 2012).

Facilitating enhanced financialisation of farmland, through actors such as private equity firms, potentially alters how the value of farm land is understood, and how farming is conducted in Australia. In addition, by encouraging farmers to attract private investment, the White Paper proposes that farmers consider business models which cede full ownership of the farm to external investors, such as joint ventures, leasing arrangements, and share-farming (Commonwealth of Australia 2015). The latter approach could potentially allow smaller farmers to develop greater scale by pooling resources, while still retaining a share in the decision making, with fellow farmers. This could potentially allow farming collectives to develop and for farmers to remain in the industry, without ceding control to private investors. However, in other scenarios described by the White Paper, such as joint ventures featuring multiple investors, or farmers selling land to external or offshore investors, to remain on their property as farm managers, it is conceivable that farmers will lose their autonomy and independence. Self-reliance, in this sense, is constructed in terms of the farmers' capacity to attract external funding. An independent farmer is framed as one who makes good business decisions and, in addition, can secure private sector funding.

How this influences farmers' conceptions of themselves, and their land, will be an important development to follow. Furthermore, if investors such as private equity firms are intent on short-term returns, and the involvement of these types of investors in Australian farm land increases, how will this alter land use and farming methods, and what will be the environmental implications of these changes? These are important questions, which the White Paper has not sufficiently considered in calling for greater private sector investment in Australian farming. What my analysis does reveal is that this shift reflects a subtle, yet substantial, shift in the responsibility for the farm sector from government, towards corporate interests.

Conclusion

This chapter charts the discursive shift towards the construction of the good farmer as one who is rational, entrepreneurial, and business minded, and, as the recent Agricultural Competitiveness White Paper shows, focused on reconstructing their farm businesses to attract private investment. This work builds upon governmentality research exploring the operationalisation of individualised farmers, using Dean's (1999) concept of technologies of agency.

This shift is underpinned by the rationalisation of farming, and the disconnection of the business of farming, conceptualised by policy makers as the conversion of resources into commodities, from farmers' roles as family members, community members, and custodians of the land. Policy discourses marginalise the significance of these roles, and the importance of farming to farmers' mental well-being, and instead construct the value of farming as being exclusively understood in terms of farmers' capacity to manage "the community's resources" efficiently and productively. Thus, good farming becomes associated with the farming processes which are portrayed as meeting these broader aims, such as productivist farming methods, and the attraction of private investment. The former enhances productivity, whereas the latter relieves the state, and the community, of the burden of financing agriculture. The state's role, in turn, is repositioned as an enabler of skill development (McKee 2008), by promoting programmes which assist farmers to develop the skills necessary to develop "good farming" practices and knowledge.

In this regard, structural adjustment and policies which indirectly facilitate structural adjustment, such as deregulation of the wheat export market, are conceptualised as being positive shifts which benefit the community. Whereas these policies cause farmers to leave the industry, this process is justified on the basis that these farmers are simply inefficient, "sloppy operators" who were unable to adapt to the changing environment (Financial Review 1992). Policy discourses allege that the exits of these farmers are overwhelmingly positive, as the resources which these inefficient farmers controlled are liberated to be put to more productive use by the most efficient farm managers. As a result, the efficiency of the farm sector and its contribution to the national economy is enhanced.

In addition, good farming is constructed as an attitude, whereby the good farmer is one who prioritises income maximisation through exercising their freedom and choice, and by utilising their skills and knowledge in competitive markets. The good farmer does not want protection from markets, they want to test their abilities in liberated markets and use markets to their advantage. This construction was used by policy makers to facilitate wheat export market deregulation. Politicians argued that as the good farmer wants the freedom and choice promised through wheat export market deregulation, this policy shift is in the farmers' best interests.

In a similar manner, this construction of the good farmer, as one which is able to attract private investment, is now being used by policy makers to facilitate greater farm financialisation in Australia. However, this shift has the potential to greatly undermine farmers' capacity to exert power over decision making, which could be transferred to private equity partners, for example, who may be more intent on deriving short-term profits from their investments. This represents a further shift towards private ownership in the Australian farm sector, with potentially significant implications for farming and food production.

References

Allen Consulting, (2000), *The Wheat Marketing Act 1989 – The Economic Impact of Competitive Restrictions*, Sydney: Allen Consulting Group.

Argent, N., (2005), The neoliberal seduction: Governing-at-a-distance, community development and the battle over financial services provision in Australia, *Geographical Research*, 43(1), pp. 29–39.

Burch, D., Dixon, J., and Lawrence, G., (2013), Introduction to the symposium on the changing role of supermarkets in global supply chains: From seedling to supermarket: Agrifood supply chains in transition, *Agriculture and Human Values*, 30, pp. 215–224.

Cheshire, L., and Lawrence, G., (2005), Neoliberalism, Individualisation and Community: Regional Restructuring in Australia, *Social Identities*, 11(5), pp. 435–445.

Clapp, J., (2015), Distant agricultural landscapes, *Sustain Sci*, 10, pp. 305–316.

Commonwealth of Australia, (2013), *Agricultural Competitiveness White Paper: Terms of reference*, [online] http://agwhitepaper.agriculture.gov.au/supporting-information/key-documents/terms-of-reference [Accessed December 12 2016].

Commonwealth of Australia, (2015), *Agricultural Competitiveness White Paper: Stronger farmers, stronger communities*, [online] http://agwhitepaper.agriculture.gov.au/ [Accessed June 6 2016].

Dean, M., (1999), *Governmentality: Power and Rule in Modern Society*, London: Sage Publications.

Dibden, J., Potter, C., and Cocklin, C., (2009), Contesting the neoliberal project for agriculture: Productivist and multifunctional trajectories in the European Union and Australia, *Journal of Rural Studies*, 25(3), pp. 299–308.

Fairbairn, M., (2014), 'Like gold with yield': Evolving intersections between farmland and finance, *Journal of Peasant Studies*, 41(5), pp. 777–795.

Financial Review, (1992), *The best in the bush*, August 28 1992.

Gill, F., (2011), Responsible agents: Responsibility and the changing relationship between farmers and the state. *Rural Society* 20(2): 128–141.

Gray, E., Oss-Emer, M., and Sheng, Y., (2014), *Australian agricultural productivity growth: Past reforms and future opportunities*, Australian Bureau of Agricultural and Resource Economics and Sciences, Research Report 14.2, Canberra, February.

Higgins, V., Dibden, J., and Cocklin, C., (2015), Private agri-food governance and greenhouse gas abatement: Constructing a corporate carbon economy, *Geoforum*, 66, pp. 75–84.

Higgins, V., (2002), *Constructing reform: Economic expertise and the governing of agricultural change in Australia*, New York: Nova Science Publishers.

Higgins, V., (2001), Governing the boundaries of viability: Economic expertise and the production of the 'low income farm problem' in Australia, *Sociologia Ruralis*, 41(3), pp. 358–375.

Higgins, V., and Lockie, S., (2002), Re-discovering the social: neo-liberalism and hybrid practices of governing in rural natural resource management, *Journal of Rural Studies*, 18, pp. 419–428.

Hilmer, F, Rayner, M., and Taperell, G., (1993), *National Competition Policy Review*, [online] http://ncp.ncc.gov.au/docs/National%20Competition%20Policy%20Review%20report,%20The%20Hilmer%20Report,%20August%201993.pdf [Accessed July 30 2015].

Industry Commission, (1991), *Statutory marketing arrangements for primary products*, Report No. 10, Canberra: Australian Government Publishing Service.

Irving, M., Arney, J., and Linder, B., (2000), *National Competition Policy review of the Wheat Marketing Act 1989*, Canberra: National Competition Policy – Wheat Marketing Act Review Committee.

Knopke, P., O'Donnell, V., and Shepherd, A., (2000), *Productivity growth in the Australian grains industry*, ABARE Research Report 2000.1, Canberra.

Larner, W., (2000), Post-welfarist state governance: Towards a code of social and family responsibility, *Social Politics* 7(2): 244–265.

Lawrence, G., Richards, C., and Lyons, K., (2013), Food security in Australia in an era of neoliberalism, productivism and climate change, *Journal of Rural Studies*, 29(1), pp. 30–39.

Lockie S, and Higgins, V., (2007), Roll-out neoliberalism and hybrid practices of regulation in Australian agri-environmental governance, *Journal of Rural Studies*, 23(1), pp. 1–11.

Martin, S., and Clapp, J., (2015), Finance for agriculture or agriculture for finance?, *Journal of Agrarian Change*, 15(4), pp. 549–559.

McKee, K., (2008), Transforming Scotland's public-sector housing through community ownership: The reterritorialisation of housing governance, *Space and Polity*, 12(2), pp. 183–196.

Miller, P., and Rose, N., (1990), Governing economic life, *Economy and Society*, 19(1), pp. 1–31.

Murphy, S., Burch, D., and Clapp, J., (2012), *Cereal Secrets: The world's largest grain traders and global agriculture*, [online] http://www.oxfam.org/sites/www.oxfam.org/files/rr-cereal-secrets-grain-traders-agriculture-30082012-en.pdf [Accessed October 28 2014].

Nevile, J., (1971), *The government, the economy and productivity*, in Nevile, J., Le Couteur, G., and Bell, H., (eds), Productivity, Sydney: Committee for Economic Development of Australia.

Nossal, K., and Gooday, P., (2009), *Raising productivity growth in Australian agriculture*, ABARE project 3425, Canberra: Australian Bureau of Agricultural and Resource Economics.

Nossal, K., Zhao, S., Sheng, Y., and Gunasekera, D., (2009), Productivity movements in Australian agriculture, *Australian Commodities*, 16(1), pp. 206–216.

Pritchard, B., (2005), Unpacking the neoliberal approach to regional policy: A close reading of John Freebairn's 'Economic policy for rural and regional Australia', *Geographical Research*, 43(1), pp. 103–112.

Productivity Commission, (2010), *Wheat export marketing arrangements: Report no. 51*, AGPS, Canberra.

Productivity Commission, (2005), *Trends in Australian agriculture*, Canberra: Commonwealth of Australia.

Royal Commission into Grain Storage, Handling and Transport, (1988), *Royal Commission into Grain Storage, Handling and Transport*, Canberra: AGPS.

Rural Industries Research and Development Corporation, (2007), *Drivers of structural change in Australian agriculture*, RIRDC Publication No 07/057, Canberra: Rural Industries Research and Development Corporation.

Salerno, T., (2017), Cargill's corporate growth in times of crisis: How agro-commodity traders are increasing profits in the midst of volatility, *Agriculture and Human Values*, 34, pp. 211–222.

Sippel, S., Larder, N., and Lawrence, G., (2017), Grounding the financialisation of farmland: Perspectives on financial actors as new land owners in rural Australia, *Agriculture and Human Values*, 34(2), pp. 251–265.

CHAPTER 5

Acting on Society: Quantification, Technologies of Performance, and Erasure of "the Social"

Introduction

This chapter analyses the use of quantification to make the wheat industry amenable to governing and, policy makers' use of technologies of performance to make wheat export market deregulation possible. Specifically, this assemblage of technologies, including performance objectives, cost-benefit analyses, econometric modelling, and audit, is focused on assessing and coercing the AWB, to narrow its value and focus towards maximising growers returns. These technologies helped construct wheat prices as an integral mechanism for measuring and assessing the performance of statutory wheat marketing. In addition, this contributed to the construction of wheat prices and supply chain costs as the primary data for understanding value and performance of wheat export marketing, and the maximisation of wheat prices as the central ambition of wheat marketing policy.

This is a pivotal element in the deregulation of the wheat export market. Statutory wheat marketing was not established to maximise growers' returns. Specifically, statutory wheat marketing was implemented to provide growers with security and stability, and to support growers collectively by redistributing returns evenly. Thus, problematising the concept of equity in industry policy and constructing efficiency and the maximisation of wheat prices as the central focuses of wheat export market policy marginalised the AWB, and the value that growers accorded to statutory marketing.

I commence this chapter by drawing on policy discourses which construct efficiency as the primary ambition of agricultural policy. This is

P. O'Keeffe, *Making Markets in Australian Agriculture*,
https://doi.org/10.1007/978-981-13-3519-8_5

made possible through the separation of equity and efficiency, and the marginalisation of equity as a credible policy objective. Next, I turn to the technologies which make the maximisation of efficiency possible: policy objectives, audit, cost-benefit analysis, and econometric modelling. I argue that these technologies shape how the wheat industry is understood and are used to develop legitimate, quantitative knowledge that demonstrates the superiority of liberalised markets. In addition, I suggest that this provides the means of governing through organisations such as the AWB. Following this, I turn to the discursive techniques employed by policy makers to de-legitimise and erase the social world. This marginalises opposition to deregulation, on the one hand, while also obscuring the negative social and environmental consequences of deregulation.

Prioritising Efficiency

In 1974, the Commonwealth Government established the IAC to "improve the efficiency with which the community's productive resources are used" (Industries Assistance Commission Act 1973, section 22.1.a). This repositioned agricultural policy, which became focused on improving the productive and efficient management of the "community's productive resources". Rather than protecting farmers, the IAC sought to protect the community's right for its productive resources to be managed with maximum efficiency, contributing to national economic performance. Intended or otherwise, this diminishes the position of farmers. Farmers' claims for security and stability, central to the collectivism which marked agricultural policy following the Second World War, are substantially weakened, as their role is recast in terms of their capacity to manage the community's resources productively and efficiently.

The IAC (1983, p. 6) reiterated this emphasis on efficiency in its 1983 review of the Australian wheat industry, stating:

> Policy should have regard primarily for the efficiency with which the community's resources are used…this view is not based on a belief that other community objectives are any less important but rather on the belief that other objectives can be achieved more directly, and at less cost, through instruments which are not part of industry assistance policy.

This statement separates efficiency and equity as policy objectives, reflecting the perspective of mainstream policy documents addressing wheat industry regulation from the 1980s onwards. This portrays efficiency maximisation as a credible objective of good industry policy, whereas equity is framed as antagonistic towards this ambition. Within this policy framework, the security, stability, and equalising measures afforded to farmers by statutory wheat marketing are cast as policy objectives which lack credibility and undermine efficiency.

This is evident in the report of the Royal Commission into Grain Handling, Storage and Transport, initiated by the Hawke government in 1986. The Royal Commission urged state governments to restructure grain handling organisations, "to ensure that the agencies are freed of social obligations" and act as commercial entities (1988, p. 46). As stated in the Royal Commission report, social obligations "can impose costs that would not usually be incurred in a purely commercial environment and these costs are ultimately borne by growers" (Royal Commission into Grain Handling, Storage and Transport 1988, p. 46). Separating equity and efficiency allows policy makers to argue that industry policy should prioritise efficiency, while claiming equity can be most effectively met through alternate policy instruments. The success of these arguments fundamentally shaped the debate over wheat industry regulation. Industry policy was guided by the question of how to maximise efficiency, with the solution presumed to be through the development of competition.

Making Reality Governable: Performance Objectives, Audit, and Cost-Benefit Analysis

In this section, I outline how assemblages of technologies are employed by governance regimes to act upon the ambition of maximising efficiency. In particular, I show that these technologies have been used to govern the performance of the AWB, first, by constructing the maximisation of growers' returns as the primary focus of wheat marketing policy, and second, to demonstrate liberalised markets as the most effective approach for maximising growers' returns. In turn, I outline how performance objectives, audit, cost-benefit analysis, and econometric modelling have been used to make this reality governable.

Creating Performance Objectives

In the early 1980s, policy makers' problematisations of government intervention into markets and industries gained mainstream appeal. This shift was underpinned by the rationality of markets as efficient, powerful mechanisms for improving economic performance. The AWB's existence prevented this rationality from being extended to the Australian wheat industry. Thus, I suggest, policy makers commenced a *project* of wheat market deregulation, by making the wheat industry amenable to this rationality. First, policy makers identified the need to measure the AWB and its performance. According to the IAC (1983, p. ii), "it is difficult to assess whether unnecessary costs are created by its involvement in activities which are not essential for it to market wheat efficiently" (IAC 1983, p. ii). Immediately, this narrows the AWB's value to its capacity to "market wheat efficiently". From this point, the IAC then raised the need to assess the performance of the AWB, stating (IAC 1983, p. 59):

> The nature of the environment in which such authorities operate also provides few benchmarks for either the authority or those outside the authority to determine whether unnecessary costs are incurred.

For policy makers (BAE 1987; IAC 1983; Victorian Government 1983), this problem is compounded by the absence of any performance objectives for the AWB. Subsequently, the IAC recommended that the AWB's primary objective should be to "maximise returns to growers from the sale of wheat without taxing domestic consumers" (IAC 1983, p. 59). This objective draws the AWB's focus towards maximising returns and incorporates the interests of consumers, as though consumers are as invested in wheat marketing as farmers. In addition, the AWB is compelled to understand and measure its value in these terms. This is significant in shaping "what matters", in relation to wheat marketing policy.

Auditing the AWB

The IAC (1983) recommended auditing to measure the AWB's performance against the objective of maximising growers' returns. This was portrayed as necessary in assisting proper assessment and accountability of the AWB. As mentioned by the IAC (1983, p. 60):

> Growers and the community are not given much opportunity to examine the affairs of the AWB. The Annual Reports and newsletters of the AWB provide little information for growers or others to assess the commercial performance of AWB.

The IAC claimed auditing is in the growers' interests. Yet, there is no indication that growers were concerned with the apparent lack of transparency or felt that auditing should be used to monitor the AWB. To address this shortcoming in relation to accountability and data, the IAC (1983, p. 60) recommended that the AWB "publish separate accounts for sales on the domestic and export markets, including separation of identifiable and joint costs", thereby allowing Members of Parliament to audit its performance. Using the technology of audit in this manner is intended to create a quantitative knowledge base, which can be used to assess the AWB's performance measures such as price premiums and supply chain costs. In addition, there is no indication of what constitutes good performance, or why performance against these indicators is preferable to measuring the value that growers ascribe to the AWB, for example. The "commercial performance of AWB" is assumed to be the measure of what matters, which all parties can agree upon, yet this subtly marginalises the AWB's worth to growers. Wheat marketing, and its broader role in Australian agriculture, is simplified, reducing the significance of other factors previously central to wheat marketing policy, such as equity and the security of farmers and farming communities.

The subsequent audits of the AWB contain no significant information that would allow farmers to interpret the AWB's performance. These documents state only that audits were completed, without providing details on the findings of these audits (Australian National Audit Office 1989, 1990, 1991, 1992, 1993, 1994). Rather than enhancing transparency and accountability, auditing the AWB provided the state with a technology to guide its performance to focus on maximising efficiency. In this regard, the construction of objectives and use of audits make the AWB calculable and therefore governable.

National Competition Policy: Quantifying Costs and Benefits

In 1993, the Keating government initiated the National Competition Policy Review, to create a national framework for developing "an open, integrated domestic market for goods and services by removing unnecessary barriers to trade and competition" (Hilmer et al. 1993, p. 361). As argued by Hilmer et al. (1993, p. 1):

> If Australia is to prosper as a nation, and maintain and improve living standards and opportunities for its people, it has no choice but to improve the productivity and international competitiveness of its firms and institutions.

This opening statement presents regulation as a barrier to national prosperity and reiterates the necessity and inevitability of economic restructuring, under the claim of raised living standards. Using Hilmer et al.'s framework, the Keating government subsequently implemented the NCP in 1995. As explained by the Productivity Commission (2005, p. iv), "A key principle of NCP is that arrangements detracting from competition should be retained only if they can be shown to be in the public interest." Analyses of policy within this framework therefore assumed that competition would confer economic benefits to society, and that in most cases, regulation restricting competition imposed costs which were detrimental to society's broader interests (Productivity Commission 2000, 2005; Irving et al. 2000). Subsequently, assessments of the value of regulation were reduced primarily to the economic costs and benefits that it generated.

This leads to the National Competition Policy Review of the Wheat Marketing Act. Announcing this review, Minister for Agriculture, Fisheries and Forestry, Warren Truss, asked the Independent Committee, chaired by Malcolm Irving, to "analyse and quantify the benefits, costs and overall effects on businesses involved in the Australian wheat industry and/or the community generally" (Irving et al. 2000, p. viii). According to this statement, to analyse, or to understand, is to quantify. Applied by Irving et al. (2000), this analysis of the quantifiable costs and benefits can be understood in terms of supply chain costs and wheat prices, as evidenced by examination of the AWB's performance in statutory markets, and the projections of econometric modelling. According to Irving et al. (2000, p. 118):

> In the Committee's view, the question of the size of the single desk price premium is pivotal, *since much of the argument for the single seller system providing a benefit appears to depend on the idea that the system provides significant additional export returns to growers* over and above what would be provided by a multiple seller system. [Emphasis added]

Yet, as this chapter demonstrates, the AWB's objective to maximise grower returns was implemented at the IAC's recommendation.

This emphasis on wheat price premiums and supply chain costs is pivotal, and is made possible through the subtle use of apparently humble and mundane technologies, such as policy objectives, audit, and cost-benefit analysis. This policy shift towards focusing on growers returns is made to appear self-evident, as common sense. Policy makers imply that it makes sense for a wheat marketing authority to focus wholly on maximising wheat prices. Auditing the AWB's performance against this objective is portrayed as the obvious actions of a responsible government, as is using cost-benefit analysis to determine if regulation outperforms market liberalisation. Yet, each of these technologies are dependent upon the construction of value as being that which can be produced and communicated through markets, liberalised or otherwise. This is a subtle shift in power away from the AWB, statutory marketing, and farmers, to the extent that the eventual deregulation of the wheat board is constructed as a common-sense decision.

Assemblages of Technologies

This emphasis on wheat price premiums and supply chain costs is pivotal. Yet it is also made possible through the subtle use of apparently humble and mundane technologies, such as policy objectives, audit, and cost-benefit analysis. This policy shift towards focusing on growers returns is made to appear self-evident, as common sense. Policy makers imply that it makes sense for a wheat marketing authority to focus wholly on maximising wheat prices. Auditing the AWB's performance against this objective is portrayed as the obvious actions of a responsible government, as is using cost-benefit analysis to determine if regulation outperforms market liberalisation. Yet, each of these technologies are dependent upon the construction of value as being that which can be produced and communicated through markets, liberalised or otherwise. This is a subtle shift in power away from the AWB, statutory marketing, and farmers, to the extent that the eventual deregulation of the wheat board is constructed as a common-sense decision. However, in conjunction with these technologies are the discursive constructions of equity, subjectivity, and the social world, which have been used in policy discourses to effectively externalise the social impacts of deregulation, and erase the significance of the social world in policy making. I now turn to these constructions and their role in making wheat export market deregulation possible.

Erasing the Social

Quantification enables policy makers to construct acceptable knowledge, to shape a simplified conception of reality. The legitimacy of this reality is dependent upon the erasure of the social world, in policy terms. The complexity of the social world interferes with policy makers' reductionist approach to agriculture in particular, where key policy questions are focused on how to best maximise the productive use of resources. The answer to this question, as constructed by policy discourses, is that this is best achieved through liberalised markets which promote allocative efficiency. This approach requires a distancing of policy makers from society. Quantification is central to creating distance, allowing the state to view society as an economy, comprising of distant and distorted, movable, and replaceable parts. In turn, quantification enables policy questions to be reduced to equations which can be solved through the manipulation of numbers which are distant of the world they are claimed to represent. Dismissing competing discourses, and marginalising the value of qualitative reflections on lived experiences and observations, is central to this task. This chapter now analyses the discursive techniques used by wheat marketing policy reviews (Productivity Commission 2010; Irving et al. 2000) to erase the value of the social world.

Subjectivity of Equity

Policy discourses frame equity as value laden, subjective, and difficult to measure and understand. For example, the Productivity Commission (2001, p. 45) claims that policy arguments based on equity are controversial and contentious, due to the fact that it is an "inherently subjective and value laden" concept. The subjectiveness of equity is viewed as creating "inconclusive, drawn-out political debates about equity or regional impacts, most of which are unknowable and unprovable" (Productivity Commission 2001, p. 3). Significantly, this questions the validity of information based upon subjective interpretation. Conversely, the Productivity Commission (2001, p. xix) claims that the arguments for efficiency are simpler and clearer, and which can be made knowable through an assessment of the costs and benefits of regulation, in relation to the projected benefits and costs of deregulation. This perception substantially affected the capacity of Irving et al. (2000), the National Competition Council (2004), and the Productivity Commission (2010), in particular, to fully appreciate the social benefits of wheat export market regulation, as experienced by farmers and their communities.

"Evidence" and Evidence

Studies such as the Irving et al. (2000) inquiry into wheat export marketing policy sought submissions from industry participants, including growers.

Irving et al. (2000) treat these submissions differently, depending on the participants' attitudes towards wheat export market deregulation. Despite it being "obvious that AWB Limited has very strong support from a majority of Australian wheat growers", Irving et al. afford greater import to growers' arguments criticising statutory wheat marketing (Irving et al. 2000, pp. 64–67). Irving et al. (2000, p. 64) summarise the perceived "benefits" of regulation using plain and passive language, presented as unsubstantiated claims:

> Price premiums are obtained from 'single desk' selling, which add to average export returns.

Following this, the perceived "disadvantages" of statutory marketing, according to growers, are written as arguments supported by critical appraisal of available evidence (Irving et al. 2000, p. 65):

> At best there is weak evidence the existence of significant price premiums over and above those attributable to quality and freight, [and]
>
> The evidence that the 'single desk' delivers cost minimisation in storage and handling is problematic.

Irving et al.'s (2000, pp. 65–67) presentation of growers' perceived disadvantages of statutory wheat marketing appears to be used to counter the benefits of regulation, as claimed by "the majority" of growers. Portrayed as an independent study, this demonstrates Irving et al.'s subtle marginalisation of growers' support for the AWB. This is further underlined in Irving et al.'s (2000, p. 6) conclusion that they were presented with no "clear, credible or unambiguous evidence" of the benefits of statutory marketing. Thus, the inquiry considered submissions from the "overwhelming" proportion of wheat growers supporting statutory marketing as lacking sufficient clarity and credibility to influence their conclusions (Irving et al. 2000, p. 82).

Similarly, ACIL Tasman (2004, p. 20) examined the impact of grain market deregulation, in relation to the deregulated barley and canola markets. In outlining its methodology, ACIL Tasman argues:

> Canvassing views from growers was not deemed necessary. As well as being time consuming and expensive, grain selling patterns and utilisation of new products provides sufficient indication of grower reactions and attitudes.

This reduces growers' capacity to communicate information to their decision making in markets. Thus, markets are constructed as the only reliable indicator of growers' preferences, values, and opinions. This construction dehumanises growers, by narrowing all that is significant about their experience of grain market deregulation to their market transactions. This asserts the primacy of markets as a communicator of credible information, while further diminishing the value of the social world.

Difficult to Document

Policy discourses around wheat industry deregulation construct qualitative data, particularly that received from growers, as "difficult to document" (ACIL Tasman 2004, p. 20; Allen Consulting 2000a, p. 22). Data pertaining to the social world is eschewed in favour of quantifiable data that, in this case, monopolises the term "evidence".

In assessing the impacts of deregulation of grain markets, ACIL Tasman consulted industry and rural media to identify concerns arising from this shift. This included largely economic matters. Additional concerns mentioned "were the social and environmental effects", which ACIL Tasman claims were mostly cited by those who opposed deregulation (ACIL Tasman 2004, p. 20). ACIL Tasman adopts two strategies in deflecting these concerns. First, the report tacitly claims that those raising the social and environmental effects of deregulation held a pre-existing bias, without considering that participants may have formed their view of deregulation based on their observations of negative social and environmental consequences of this shift. Second, ACIL Tasman (2004, pp. 20–21) dismisses the credibility of these concerns by claiming that they are "not very well articulated", "difficult to document", and therefore unsupported by evidence. This delegitimises these concerns as potential consequences of deregulation, as their articulation did not fit the ACIL Tasman's methodological framework.

This raises the question: if the negative social consequences of deregulation cannot be measured, do they exist? Rather than considering the broader social consequences of economic restructuring, Allen Consulting (2000b) focuses their assessment of "social impacts" upon economic impacts, such as changing wheat price premiums. This is a clear example of policy making narrowing the complexity of an issue, through the desire to create simple, easy-to-generate, quantifiable data. Resultantly, this undermines the potential social and environmental consequences of deregulation as legitimate concerns in the policy-making process.

Individualising the Social Consequences of Deregulation

Following the deregulation of the wheat export market in 2008, the Productivity Commission published an assessment of these reforms in 2010. This report is striking in its incapacity to absorb wheat growers' negative experiences of deregulation. As an example, the following comment attributed to a submission from the "Pike Family Trust" identifies the additional responsibilities of grain marketing as a significant social impact of deregulation:

> Although this 'brave new world' is welcomed by some it puts pressure on our business and families as there is not enough time to attend to all that one must, is required to and would like to within the farm let alone a life 'after hours'. (Pike Family Trust, sub. 18, p. 1)

Rather than identifying the important issues of increased workload, reduced leisure time, and increased pressures of families, the Productivity Commission (2010, p. 86) argues that wheat export market deregulation is necessary and that growers will adapt:

> Australian farmers, particularly grain growers, are resilient and resourceful and have a proven track record of adjusting to international market developments and domestic cost pressures (so-called declining terms of trade) by improving their productivity. This can mean short-term pain for some, but will deliver long-term gains in the form of a competitive and efficient wheat export industry.

Effectively, the supposed "short-term pain" of growers is constructed as less significant than the creation of a competitive wheat industry. Similarly, in predicting the social impacts of deregulation, Allen Consulting states (2000b, p. 47):

> It is clear that removal of the single desk will benefit some producers and disadvantage others. Those who are efficient and are able to manage risks (ie, financial risks, make appropriate choices as to how to market, chose appropriate wheat types, etc) will be the likely winners. *It is likely that those producers who are unable to appropriately manage risks will exit the industry and there will be a round of farm consolidations.* [Emphasis added]

In other words, the negative social consequences of deregulation, such as increased pressure on farm families, reduced leisure time, and, in the worst cases, farmers exiting the industry, are portrayed as the problems of individual farmers, rather than the structural change itself. Furthermore, these consequences of structural change are intentionally referred to in detached and distant language. The terms "short-term pain" and "a round of farm consolidations" do not convey the full extent of these outcomes. For farmers, exiting the industry creates a deep sense of loss, as farmers lose their connection to the land, their occupation, social networks, and status as farmers (Kuehne 2012). Policy discourses purporting to examine the social consequences of deregulation mask these negative consequences, which are framed as unfortunate, though necessary, externalities.

Conclusion

This chapter analyses the process of Australian wheat export market deregulation, by focusing on policy makers' attempts to define, measure, and assess the value of the AWB as a statutory marketing authority. Policy makers sought to address the issue of wheat industry regulation, by using quantification to reduce this complex industry to simplistic and narrow measures such as wheat prices and supply chain costs. These measures are constructed as legitimate knowledge and used to determine Australian wheat export market policy. This portrayed the Australian wheat industry in simplistic terms and in isolation from the impact of structural change upon farmers and their rural communities. Equity was constructed as a peripheral issue, and subjective information such as experiences and perspectives as untrustworthy data, with assessments of the wheat industry instead favouring knowledge produced by statistical methods such as econometric modelling.

This shift was designed to shape the Australian wheat industry by establishing performance objectives which could be quantitatively measured. In this manner, this enables governing in the wheat industry to occur through organisations such as the AWB, which had been founded on more collectivist notions of governing. The use of technologies of performance compelled the AWB to focus on its objective of "maximising growers' returns". Any activity not specifically focused on this objective was framed as evidence of inefficiency. This created a substantial problem for the AWB, which was essentially established to equalise returns among growers, not to maximise returns. In turn, growers valued the AWB for its role in marketing wheat, which relieved growers of the burdens of risk, workload, and associated stress. Yet through narrowing the AWB's purpose to focus on returns from wheat sales, policy makers constructed these roles as having little value, and as being oppositional to the more important goal of maximising industry efficiency.

Subsequently, inquiries performed by the National Competition Policy Review panel (Irving et al. 2000), the Productivity Commission (2000, 2010), and consultancy firms (ACIL Tasman 2004; Allen Consulting 2000a, b) focused on addressing the question of whether wheat export market regulation delivered net gains to farm businesses and the broader community. This research specifically focused on whether the AWB, or firms competing in a deregulated market, would be able to deliver premium prices to growers, while minimising supply chain costs. These studies are portrayed as disinterested investigations. However, this research marginalises growers' attempts to participate, through portraying submissions based on experiences and observations gleaned through working in the industry as unreliable, biased, and not supported by "evidence". Conversely, this research considers statistical data, particularly generated through economic modelling, as credible, neutral, and disinterested representations of reality. In this regard, policy makers' use of quantification and technologies of government depoliticises the decision to deregulate the Australian wheat export market. This is instead portrayed as the actions of a responsible government, based on an unbiased representation of reality.

Importantly, this shift also subtly transfers importance from "value" to "performance". The performance of statutory wheat marketing supplants the value that wheat growers ascribe to this regulation. Research such as that conducted by Irving et al. (2000) and the Productivity Commission (2000, 2010) focuses on measuring performance: of the economy, of the industry, and of the AWB. These objects, such as the economy and the wheat market, are constructed as real, performing entities. On the other hand, the social world is erased. The relationships between farmers, their families, family histories, land, and communities are overlooked, and considered as ancillary concerns which simply distract attention from the central purpose of wheat marketing policy, which is to maximise industry efficiency.

Quantification enables this process to occur, by constructing a reality which reflects a simplistic, reductionist wheat industry. Technologies such as audit and assessment are applied by policy makers to act upon this constructed reality, to advance the rationale of markets, efficiency, and competition.

References

ACIL Tasman, (2004), *A review of the NCP Grain Market Reforms: NCC occasional series*, Prepared by ACIL Tasman, Canberra: Ausinfo.

Allen Consulting, (2000a), *The Wheat Marketing Act 1989 – The Economic Impact of Competitive Restrictions*, Sydney: Allen Consulting Group.

Allen Consulting, (2000b), *The Wheat Marketing Act 1989 – The Social Impact of Competitive Restrictions*, Sydney: Allen Consulting Group.

Australian National Audit Office, (1989), *Report on the audit of the Australian Wheat Board, 1988–1989*, Canberra: Australian Government Publishing Service.

Australian National Audit Office, (1990), *Report on the audit of the Australian Wheat Board, 1989–1990*, Canberra: Australian Government Publishing Service.

Australian National Audit Office, (1991), *Report on the audit of the Australian Wheat Board, 1990–1991*, Canberra: Australian Government Publishing Service.

Australian National Audit Office, (1992), *Report on the audit of the Australian Wheat Board, 1991–1992*, Canberra: Australian Government Publishing Service.

Australian National Audit Office, (1993), *Report on the audit of the Australian Wheat Board, 1992–1993*, Canberra: Australian Government Publishing Service.

Australian National Audit Office, (1994), *Report on the audit of the Australian Wheat Board, 1993–1994*, Canberra: Australian Government Publishing Service.

Bureau of Agricultural Economics, (1987), *Wheat marketing and assistance*, Canberra: Australian Government Publishing Service.

Hilmer, F, Rayner, M., and Taperell, G., (1993), *National Competition Policy Review*, [online] http://ncp.ncc.gov.au/docs/National%20Competition%20Policy%20Review%20report,%20The%20Hilmer%20Report,%20August%201993.pdf [Accessed July 30 2015].

Industries Assistance Act 1973 (Commonwealth).

Industry Assistance Commission, (1983), *The Wheat Industry*, Report No. 329, AGPS, Canberra.

Irving, M., Arney, J., and Linder, B., (2000), *National Competition Policy review of the Wheat Marketing Act 1989*, Canberra: National Competition Policy – Wheat Marketing Act Review Committee.

Kuehne, G., (2012), My Decision to Sell the Family Farm, *Agriculture and Human Values* 30(2), pp. 1–11.

National Competition Council, (2004), *NCC Occasional Series: A review of the NCP grain market reforms*, AusInfo, Canberra.

Productivity Commission, (2010), *Wheat export marketing arrangements: Report no. 51*, AGPS, Canberra.

Productivity Commission, (2005), *Review of National Competition Policy reforms: Productivity Commission Inquiry Report*, No. 33, 28 February 2005, Commonwealth of Australia, Canberra.
Productivity Commission, (2001), *Structural Adjustment – Key policy issues, Commission Research Paper*, AusInfo, Canberra.
Productivity Commission, (2000), *Single-desk marketing: Assessing the economic arguments*, [online] http://www.pc.gov.au/research/supporting/single-desk [Accessed December 20 2016].
Royal Commission into Grain Storage, Handling and Transport, (1988), *Royal Commission into Grain Storage, Handling and Transport*, Canberra: AGPS.

CHAPTER 6

Freedom and Choice? Legitimising Concentration in Deregulated Agricultural Markets

Introduction

In this chapter, I analyse the development of the wheat export market following deregulation. My research shows that the industry has become increasingly characterised by regional monopolies and oligopolies. Firms have used mergers and acquisitions to develop and protect their positions in regional markets in particular. This has undermined growers' power in local markets, and not delivered the freedom, empowerment, and choice promised by policy makers.

I analyse policy makers' use of contestability theory as a foundation for interpreting the development of competition in the Australian wheat export market. I argue this theory is limited and is incorrectly applied in the case of the Australian wheat export market. Using a generalised conception of this theory, policy makers have assumed that removing regulatory restrictions on market entry is sufficient to create contestable markets. However, a key condition of contestability theory is that market entry does not incur a cost for potential market entrants. Yet to develop a substantial share of the wheat export market, firms must invest in supply chain infrastructure. These investments, including mergers or acquisitions with incumbent firms that control infrastructure, or through new investments, incur a significant cost. This cost represents a significant barrier to entry. Thus, markets may ostensibly be "free", in the sense that there are no regulations restricting firms' access; however, these markets are not necessarily open to contestation. As I suggest in this chapter, deregulated

P. O'Keeffe, *Making Markets in Australian Agriculture*,
https://doi.org/10.1007/978-981-13-3519-8_6

Australian wheat markets are only accessible to well-capitalised, transnational firms. The concentrated nature of these markets has meant that growers, in some cases, have a limited number of buyers for their wheat at their local receival site. As such, for many small- to medium-sized growers in particular, this policy shift has caused feelings of disempowerment and disenfranchisement.

What Is a Competitive Market?

In parliamentary debates which foreshadowed the eventual deregulation of the wheat export market in 2008, the Minister for Agriculture, Tony Burke, said that deregulation was essential for the industry and, most importantly, would enable growers to maximise their returns on production (Grattan 2008, Commonwealth of Australia 2008a). Members of Parliament and Senators from both major political parties forecasted a deregulated market place featuring numerous grain traders competing for farmers' wheat (Commonwealth of Australia 2008a, b). This was claimed to result in a market where growers were empowered, with multiple grain traders competing for growers' wheat.

Thus, wheat export market liberalisation was framed as a policy mechanism for expanding growers' choice, through the presence of numerous traders in the market. However, this description of a competitive market differs from how competition has been conceived in policy documents. For example, the Productivity Commission (2005, p. 286) argues that:

> ...increasing concentration in the local economy has been a desirable outcome of trade liberalisation, rather than a new problem which competition policy must address. That is, increased international competition has served to drive out much inefficient small scale and fragmented production.

According to this depiction, competition is less about the numbers of market participants, and more concerned with maximising efficiency, which is achieved through consolidated markets featuring large-scale, presumably efficient firms. This leads to the importance of contestability theory in driving Australian competition policy in recent decades, which posits that markets are competitive, provided that there are no barriers to entry. Specifically, contestability theory holds when firms incur no market entry costs, the market is susceptible to "hit-and-run" entry, and firms exiting the market incur no sunk costs (Shepherd 1984).

Contestability theory, emphasising potential, rather than actual competition, informed policy from 1988, when the Industries Assistance Commission (1988) argued for deregulation of wheat exports, to 2008, when the Rudd government's policy shift was implemented. For example, Hilmer et al. (1993, p. 3) claim that "competition between a few large firms may provide more economic benefit...due to economies of scale and scope." Hilmer et al. (1993, p. 2) draw upon Dennis (1977) to define competition as the "striving or potential striving of two or more persons or against one another for the same or related objects". Referring to Baumol (1982), Hilmer et al. (1993, p. 2) explain that:

> Recent work suggests that the real likelihood of competition occurring (potential striving) can have a similar effect on the performance of a firm as actual striving. Thus, a market which is highly open to potential rivals – known as a highly 'contestable' market – may be of similar efficiency as a market with actual head-to-head competition.

Policy documents informing Australian wheat industry policy share this interpretation, viewing competition in terms of the contestability of the market, rather than the number of firms in that market. Yet, these policy documents do not refer in detail to the three conditions of a contestable market: that entry of new firms is costless, the market is susceptible to hit-and-run entry (firms can enter the market, gain market share through reduced prices, for example, and exit the market before the incumbent firms are able to react), and entry is reversible, meaning that firms can exit the market without incurring sunk costs (Shepherd 1984, 1995). This raises the question of how contestability theory is used in policy, and whether contestability theory can be applied to the Australian wheat export market.

Contestability Theory

Contestability theory was developed by Baumol (1982), who framed this theory as a revolutionary idea in economics and industrial organisation. Baumol outlined contestability in his 1982 address to the American Economic Association, entitled "An uprising in the theory of industry structure", as a theory which intended to provide a flexible and applicable "benchmark for desirable industrial organisation" (Baumol 1982, p. 2). Contestability theory enabled policy makers to operationalise the concepts developed by Bork and Demsetz, particularly the notions of firm size and economic efficiency, and apply these ideas to competition policy. "Desirable industrial organisation", in this context, is considered to be that which maximises the efficient and productive use of resources. According to contestability theory, oligopolies and regional monopolies that typically characterise agricultural markets are not necessarily reflections of market failure (Baumol 1982; Baumol et al. 1983). Provided there are no barriers preventing market entry, such as prohibitive entry costs or regulatory barriers, these market structures are efficient and competitive—in principle and demonstrably (Baumol 1982; Baumol et al. 1983; Davies 1986). Conversely, contestability scholars argue that market efficiency is weakened by government regulation restricting entry to these markets (Baumol 1982; Baumol et al. 1983; Davies 1986). Despite claims that contestability theory lacks rigorous empirical support and is based upon questionable assumptions of markets and firm behaviour (Dasgupta and Stiglitz 1988; Stiglitz et al. 1987; Shepherd 1984, 1995; Brock 1983), this theory influenced antitrust, competition, and regulatory policy within the United States and Australia (Schwartz and Reynolds 1983; Weitzman 1983; Dasgupta and Stiglitz 1988; Davies and Lee 1988; Blaug 2001; Davidson 2012; Jones 2012).

Critics of contestability have argued that this theory has been used by policy makers to justify market concentration and to excuse market failures (Blaug 2001). Certainly, proponents of contestability theory share Bork's concern for antitrust regulation as an impairment upon the growth of firms. For example, Baumol and Willig (1986, p. 10) argue:

> We reject with equal conviction the position of those who hold that mere large size of a firm means that it must serve the economy badly, that high concentration ratios are sufficient to justify governmental restrictions upon the structure or conduct of an industry.

According to Baumol and Willig (1986, p. 10), the undesirable consequences potentially resulting from concentrated markets would be mitigated by market contestability. They argue that incumbent firms will experience potential competition, as they would actual competition, and that the three key conditions of contestability (discussed above) can be met. Baumol and Willig's argument presumes that firms will both act to maximise their efficiency and deliver gains from efficiency improvements to their consumers. Furthermore, this implies consumers have substantially more power than firms in this scenario. Each of these assumptions is questionable, yet is central to contestability theory.

Contestability theory fits with the emerging neoliberalism of the 1970s and 1980s, offering an excuse for neoclassical economists as to why real-world markets may not reflect perfect competition and not warrant state intervention. Contestability redefines the notion of competition to contend that oligopolistic or monopolistic markets can be competitive and efficient market structures. However, despite the limitations of this theory and the lack of empirical evidence which supports its assertions, policy makers have applied this theory to a myriad of policy areas in Australia. The uncritical application of contestability theory to wheat market policy, resulting in concentrated wheat export markets that are not contestable, is particularly problematic.

Applying Contestability Theory to Policy

Despite its flaws and limitations, contestability theory has been applied uncritically in policy. The supporting policy documents do not explain contestability theory nor, for that matter, clearly define competition. Rather, arguments for the dismantling of regulatory barriers are prosecuted on the basis that this leads to "more competition" or "greater contestability". Policy documents do not elaborate upon these concepts or critically reflect upon the viability and applicability of contestability theory. Articles by Baumol (Baumol 1982; Baumol et al. 1983; Baumol and Willig 1986) are given ceremonious citations, accompanied by a brief description of the idea of contestability theory. However, these policy documents do not refer to the conditions required for contestability theory to hold. This implies that contestability has been used as an idea, without condition. Thus, policy makers have simplified contestability, portraying the removal of barriers to entry as sufficient to create a contestable market.

Through applying contestability theory in this manner, policy makers rely upon the robustness of a small group of studies (Baumol 1982; Baumol et al. 1983; Baumol and Willig 1986) that do not clearly support the applicability of contestability theory to real-world markets. The limitations and weaknesses of this theory are not addressed by policy makers, despite the considerable criticism that contestability theory has attracted. Policy makers thus avoid engaging with criticisms presented by Shepherd (1984, 1995), Dasgupta and Stiglitz (1988), among others. Contestability theory is represented as fitting the overarching hope for deregulation in policy documents, which is to maximise market and industry efficiency. Policy documents presume the presence of large firms in markets will maximise efficiency, as larger firms are believed to be inherently more efficient, due to their size and scope. For example, the Productivity Commission (2005, p. 286) argues that:

> ...increasing concentration in the local economy has been a desirable outcome of trade liberalisation, rather than a new problem which competition policy must address. That is, increased international competition has served to drive out much inefficient small scale and fragmented production.

Thus, the oligopsonistic market structure fits policy makers' ambition to maximise efficiency, and such a market is claimed to be competitive as it is contestable. Yet, policy documents apply a shallow, uncritical interpretation of contestability, with the sole condition being that external firms are not prevented from entering the market by regulatory barriers.

Wheat Export Market Contestability

In the five years following the deregulation of the wheat export market, the initial growth in competition, in terms of real numbers of market participants, gradually dissipated. In particular, external firms used mergers and acquisitions as a strategy for gaining entry to this recently deregulated market. As a result, by 2013, the market had become highly concentrated in many regions, as indicated by Table 6.1.

As shown in Table 6.1, markets have become increasingly consolidated, particularly at the state level. Table 6.1 uses the Herfindahl-Hirschman Index (HHI) as a measure of concentration in the Australian wheat export

Table 6.1 Herfindahl-Hirschman Index (HHI) figures for the Australian export wheat industry (2011–2013)

Year	*Australia*	*QLD*	*NSW*	*VIC*	*SA*	*WA*
2011–2012	1278	1238.06	3271.96	1595.83	1578.41	2122.34
2012–2013	1403	2135.35	3811.07	1546.61	2717.49	1944.39

Data compiled from NSW Farmers (2014) and Grain Producers Australia (2013). HHI figures are calculated by adding the squares of the market share of each company in that market (Murphy 2006, p. 13). A market with one company will have an HHI figure of 10,000, whereas a market with 100 companies each with a 1% market share will have an HHI figure of 100

Table 6.2 Bulk handling companies' regional control of storage and handling, ports, and export markets

Supply chain segment	*CBH (WA)*	*GrainCorp (eastern Australia)*	*Viterra (SA)*
Market Share: up-country	Receives and stores approximately 90% of WA's grain	Handles approximately 75% of east coast grain	80% market share of SA up-country grain storage
Market Share: port throughput (%)	100	80–90	100
Market Share: export tonnage	48% WA bulk exports (2012–2013)	28% eastern Australian exports (2012–2013)	46% SA exports (2012–2013)

Adapted from Stretch, Carter, and Kingwell (2014, p. 11)

market and regional markets. The latter is most relevant as the distance between markets prevents a wheat grower in Western Australia from transporting and selling wheat to a marketer operating in Victoria, for example. HHI figures between 1000 and 1800 indicate moderate concentration, whereas figures exceeding 1800 reflect a highly concentrated market (Murphy 2006, p. 13). As shown in Table 6.2, in 2012–2013, every regional market except Victoria exceeded this figure. O'Keeffe and Neave (2017) interviewed wheat growers in the Australian State of Victoria, who suggested that this level of concentration is even greater at the local level, where only small numbers of buyers operate at local grain storage facilities.

Policy makers will argue that, despite the market concentration evident in Table 6.2, markets are still competitive as there are minimal regulatory barriers preventing firms from entering these markets. However, the threat of potential competition is tempered by the relationship between infrastructure ownership in the wheat supply chain and wheat export market share. Firms recognise that to enter wheat markets in a substantial way, they must own infrastructure. As Yasushi Takahashi, Mitsui Australia Managing Director, states "for us to be a competitive and attractive supplier of wheat and grain…we will have to make some meaningful investments in ports, rails and silos" (The Australian 2014). Similarly, Olam Australia Chief Executive, Bob Dall'Alba, says "unless you're one of the large quasi-monopoly holders of assets, you're marginalised in the business, and therefore you need to keep investing in port and other infrastructure" (Jasper 2014).

Firms respond to this relationship by focusing their wheat marketing efforts in areas where they control infrastructure. This has led to consolidated markets, particularly within the different wheat growing regions, which have been typically dominated by firms which control infrastructure previously managed by the state-based grain handling authorities. As Table 6.2 shows, the privatisation of this vertically integrated infrastructure has essentially allowed firms such as GrainCorp and Viterra to dominate the wheat export markets in the regions where they control storage, handling, and port facilities. External firms have been reluctant to markets where they have no control of supply chain infrastructure. Firms who do not own infrastructure in South Australia, for example, where Viterra controls 100% of the wheat supply chain, must utilise Viterra's storage facilities and ports. In doing so, this firm would effectively be providing financial support to a competitor in the wheat export market. Thus, firms which do control infrastructure are at a significant commercial advantage: their competitors in the wheat export market are also the clients of their supply chain businesses.

This presents a significant restriction upon competition. Whereas there may be no substantial regulatory barriers preventing market entry, firms are reluctant to invest in marketing operations unless they have some level of control within the wheat supply chain. This has resulted in firms implementing numerous mergers with, and acquisitions of, firms which do control supply chain infrastructure and investing in new facilities. However, as I argue in the next section, rather than reflecting the developing competition in the market, this activity is further reducing the number of smaller market participants. In addition, as these investments require substantial capital, these investments are limited to transnational agri-business firms. Thus, as I show, meaningful entry into the deregulated wheat export market is effectively restricted to those agri-business firms capable of sourcing the capital required to make investments in supply chain infrastructure.

Recent Investment in the Australian Wheat Export Industry

The recent investment in the wheat export market could be interpreted as an indicator of the success of the liberalisation of this market. However, as I suggest, this is more reflective of the increasing consolidation of ownership and capital within this industry. Table 6.3 includes the major investments occurring in this time and explains which supply chain segments are central to these investments. These details are significant, as firms

Table 6.3 Recent company investments in Australian wheat storage, ports, and marketing

Company	*Investment*	*State*	*Storage*	*Ports*	*Trading*
Bunge Limited	Developing new ports at Bunbury and Geelong (Sprague 2014)	WA, VIC		X	
Asciano	Formed JV with Itochu, Albany Bulk Handling (Thompson 2012)	WA		X	
Mitsui	Acquired 25% of Plum Grove (Tasker 2012)	WA			X
Louis Dreyfus	Acquired fertiliser company Ravensdown	WA	X	X	
Louis Dreyfus	Formed JV with Mountain Industries (parent Asciano)	NSW	X		
Glencore	Acquired Viterra in 2012	SA	X	X	X
Glencore	32.5% interest in Newcastle Agri Terminal JV with CBH and Olam Grains Australia (Thompson 2015)	NSW		X	
Olam Intl.	Has a 32.5% interest in Newcastle Agri Terminal with Glencore (Thompson 2015)	NSW		X	
Mitsubishi	Acquired 80% equity in Olam Grains Australia (World Grain 2014a) and 20% ownership of Olam International (Terazono 2016)	NSW		X	
Nidera	Acquired controlling share in PentAG Commodities	QLD			X
Noble Group	Quattro Ports JV with Qube Holdings	NSW		X	
COFCO	Acquired Noble Group and Nidera (World Grain 2014b, 2016)	Intl.		X	X
Sumitomo	Acquired Emerald Grain in 2014 and owns Melbourne Port facilities, through 2005 acquisition of ABA (Binsted 2014)	VIC	X	X	X
CHS	Acquired 50% share in Broadbent Grain, which has since been renamed CHS Broadbent (Broadbent Grain 2014)	NSW	X		
CHS	50:50 owner of Agfarm with Ruralco	Aus.			X
ADM	Acquired Toepfer international (ADM 2014)	Intl.			X
ADM	Attempted to acquire GrainCorp.	NSW, VIC, QLD	X	X	X
Cargill	Acquired AWB GrainFlow	NSW, VIC, QLD	X		X
Wilmar Intl.	Developed 50:50 JV with Gavilon, Wilmar Gavilon (Queensland Bulk Terminals 2017)	QLD		X	X
Marubeni	Parent company of Gavilon (Iwata 2015)	QLD		X	X

recognise that if they are to develop a share of the export market, then it is essential to own supply chain infrastructure (O'Keeffe 2016b, 2017). These investments are driven by firms' desire to develop power through developing resources and reduce dependencies upon the incumbent bulk handling companies. Firms such as GrainCorp and Viterra control bottleneck infrastructure such as ports, which are accessed by whole wheat growing regions. Competing grain traders are in the position where they must access these facilities, thereby contributing to the financial resources of their export competitors, or develop their own infrastructure.

As Table 6.3 shows, this investment has come from transnational agrifood firms, which in many cases have sought to acquire or merge with firms controlling existing wheat industry infrastructure such as ports and storage facilities. For example, Glencore acquired Viterra, which itself had acquired the privatised bulk handler ABB in 2009 (Sydney Morning Herald 2012). This provided Glencore with ownership of the entire wheat supply chain in South Australia, one of Australia's largest wheat-producing states (Stretch et al. 2014). In addition, a number of investments in smaller firms have been completed. Mitsui and Co, Salim Group, and Seaboard have each acquired 25% of West Australian trading company Plum Grove (Tasker 2012). Louis Dreyfus acquired fertiliser company Ravensdown, giving the firms access to storage and a port berth in Western Australia (Cattle and Bettles 2013). US cooperative CHS acquired a 50% stake in NSW wheat storage and handling company Broadbent Grain (Heard and Marshall 2014). Dutch firm Nidera acquired a controlling stake in PentAG Commodities, though was subsequently acquired by Chinese firm COFCO (Marshall 2015). These investments increase the control of the large, transnational firms within the Australian wheat industry, while reducing the number of smaller market participants. However, despite this consolidation of supply chain ownership, concentration is not necessarily framed as a problem. Australian policy discourses have cultivated the idea that competition is achieved by reducing regulatory barriers to market entry and exposing incumbents to the threat of potential competition. Market structure and concentration is portrayed as a secondary concern, which, if anything, could reflect the efficiency of the market due to the presence of efficient, large-scale firms.

This reflects the discursive construction of large firms as essential, efficiency maximising firms as desirable market participants. Not only is concentration of ownership framed as inevitable; the success of the dominant firms, who are portrayed as those best equipped to maximise efficiency and productivity, is dependent upon their capacity to increase scale and access to resources. These examples normalise firms' increased scale and market position, and seek to portray this as not only essential for the success of firms and the industry but also important in the increased efficiency and productivity of resource use, which, as this narrative asserts, is ultimately beneficial for consumers.

Whereas this present level of investment in the Australian wheat industry could be portrayed as evidence of increasing industry competition, the effect could be the opposite. The competition that *is* occurring is between oligopolistic transnational firms, which are acquiring smaller upstream firms to develop their power within the industry. As firms need to control storage, handling, and port facilities to be competitive, and control is achieved through investments that only the very large well-capitalised global firms appear capable of making, the contestability of the wheat export market may actually be very limited. As the data in Table 6.3 shows, it is only the well-capitalised, transnational firms which are executing these investments, ultimately removing the smaller participants from the market. Thus, these markets are not contestable; at least, according to contestability theory. In particular, market entry is dependent upon the significant investment of capital, which may not be retrievable, should the firm exit the market. Furthermore, market entry is effectively restricted to firms capable of making these investments.

The Effects of Consolidated Regional Markets upon Growers

As O'Keeffe and Neave (2017) and Baker (2018) have shown, rather than experiencing the promised benefits of choice and freedom, wheat growers have had to negotiate consolidated wheat markets. Consequently, the dismantling of the favoured AWB and the subsequent shift in industry control towards transnational agri-business firms have created feelings of disillusionment and disenfranchisement among some growers, who feel deceived:

> At the time, I was very naïve. I just thought, "But it's got to be good for us. Because there is going to be competition"...All of it is to get control away from farmers. (O'Keeffe and Neave 2017, p. 15)

In addition, growers suggest that the agri-business firms who now control the industry were integral to initiating the deregulation of the wheat export market:

> People who set out to dismember the wheat board are those international players who now own, not only the people who take in the grain and store it, but they also own the grain. So they own it all. The supply chain, right through. (O'Keeffe and Neave 2017, p. 12)

First, these findings suggest that the liberalisation of the wheat export market has failed to meet policy makers' claims of increased power and choice for wheat growers. However, drawing upon my analysis of the discursive construction of firms and markets in Chap. 3, I argue that instead of being centred on supporting growers' needs, this policy shift was primarily focused on permitting firms to enter this market, in the presumption that real and potential competition would compel firms to maximise efficiency, thereby raising the economic performance of the industry. As I argue in Chap. 4, wheat export market deregulation is justified through policy discourses as being in the interests of the individual, entrepreneurial "good farmer", while on the other hand, contributing to the exits of farmers who are unable or unwilling to adapt to the deregulated wheat market. Yet as the farmers' quotes highlighted in O'Keeffe and Neave (2017) demonstrate, this policy shift has also precipitated a shift in control within the industry, away from farmers and towards agri-business firms. This shift in control, according to these growers, is associated with consolidating ownership, on the one hand, and influence over policy making, on the other. As I show in this chapter, these strategies are designed to enhance the power and control of these firms, which is at odds with the discursive construction of firms as lacking power over markets and policy makers.

Resistance

For many farmers, there is little option to the productivist approach to farming, and the continued expansion that this approach encourages. However, as the following research suggests, there are farmers who are rejecting productivism and seeking to nurture their relationship to the land, their customers, and their communities. The following edited excerpt is taken from O'Keeffe (2016a), which is a study of 23 wheat farmers based in the wheat growing region of western Wimmera, in Victoria, Australia:

> While many included in this study lamented the social impacts of increasing farm sizes and the declining farming populations in the local area, very few

challenged the belief that the security of their farming operation was directly related to their capacity to continually expand in scope, and in doing so, increase their productivity and efficiency. Participants felt that in a climate where farming populations were declining, the smaller properties were most vulnerable to becoming "uneconomic". This compelled farmers to continually increase the size of their properties. This pressure is explained by one farmer, who was now working with his brother and nephew, on a farm which had been managed by their father:

As long as you are average or above, then there's a fair chance that your business is sustainable. As long as you are average or above, the pressure is not on you.

Throughout the interviews, participants expressed that this property growth was financed by increasing debts, which caused considerable stress for some. In addition, this increasing scale and scope of farming ensured the majority of farm businesses involved the active participation of numerous family members.

The perceived compulsion to continually expand the size of their properties was viewed by some participants in this research as a trap which they could not avoid. However, for other farmers, this contributed to a distancing from themselves, and the practice of "actually farming". They explained what they perceived as a diminished connection between the farmer, their market and the people that consume their food, stating that farming had become a "clinical business". This farming entity was comprised of a husband and wife, who felt that productivist farming was the antithesis of their own approach to agriculture, which prioritised their sense of connection with the land, and their own consideration of themselves as custodians of that land. As one member of this partnership stated:

It's almost as if (farming) has lost its soul…It's just become another factory business. And that is fine if that is what you want. But that is not what we want. And that comes back to having long links with the land, and wanting to improve the land and not make it worse.

For a small percentage of participants, rejecting the productivist paradigm was an avenue through which they could increase the viability of their farm. This was achieved through diversification, and a focus on developing products which met the needs of consumers seeking free-range, or chemical-free produce, and through the use of weed and pest control methods which did not require chemical inputs. Furthermore, for some farmers, the decision to limit their expansion allowed them to retain a greater connection with their land. This is indicated through an exchange involving a husband and wife partnership:

> ...this connection that you build up with the land develops when you do the work yourself.
>
> It's sort of like being a caretaker. And the older I get the more I feel that. That it's not ours forever, so therefore it's not ours to rape and pillage and get everything you can out of it for a season...And I think that this is something that has been lost. Personally I think it's mainly a financial thing. People just get into so much debt...that [they] have no choice but to drain the land for all it is worth.
>
> These comments highlight the importance that some farmers attribute to the connection that they have with their land, and how this may influence their decisions. However, this quote also alludes to the role that financial pressure plays in causing some farmers to farm their land with an intensity which may not be environmentally sustainable. Yet for these farmers, moving away from this intensive approach to farming is a way of expanding their business without expanding the size of their operation. This directly questions the prevailing belief that viability is directly related to the continual expansion of property size.

This suggests that despite the prevailing agricultural and economic discourse in Australia and its emphasis on creating economically rational, individual farmers, this path is not determined. Resistance is possible and is being practised by farmers who are using creative and innovative ways to practise farming in accordance with their own values and their own sense of what matters.

Conclusion

This chapter analyses policy makers' application of contestability theory to Australian wheat export market policy. Competitive markets, according to policy discourses around competition policy, are those which are contestable. That is, incumbent firms may appear to control the market in the present; however, the potential threat posed by new market entrants will compel firms to maximise efficiencies. Thus, a market is contestable, provided there are no barriers preventing firms from entering that market. Despite the withdrawal of regulatory restrictions on entry into the wheat export market, the market does not meet the key conditions of contestability theory, as firms seeking to develop their market position must control supply chain infrastructure.

Firms' share of the wheat export market is directly related to their control of wheat industry infrastructure, such as ports and storage facilities. Therefore, the capacity of new firms to challenge market incumbents is limited by their ownership of supply chain infrastructure or, rather, lack thereof. To develop a substantial share of the market, firms must invest in infrastructure, either through new investments, or by mergers and acquisitions with incumbent firms that control infrastructure. These investments require significant capital. Meaningful entry into the wheat export market therefore incurs considerable expense. As a result, these markets are not contestable, at least in relation to the conditions required for contestability theory to hold.

Consequently, the wheat export market has become highly concentrated at the regional level, dominated by vertically integrated firms which control the supply chains in these areas. For growers, the promises of choice and freedom have not come to fruition. Policy makers have succeeded in facilitating the entry of large, well-capitalised firms into the market. This has undoubtedly stimulated investment in the sector. This also reflects policy makers' preference for large firms, in the presumption that these actors will use their superior efficiencies and managerial capacities to maximise the productive use of the nation's resources and to enhance the productivity and efficiency of the industry. In this regard, it appears that arguments purporting to implement wheat export market deregulation, in the interests of growers, are disingenuous. First, growers' consistently and steadfastly opposed this deregulation of the market. Second, the claimed benefits of this shift have not come to fruition. Third, I argue that this policy shift was instead designed to maximise the industry's efficiency and productivity through facilitating entry of large, well-capitalised firms and also stimulating structural adjustment through farmer exits.

References

ADM, (2014), *ADM completes acquisition of remaining stake of Toepfer International*, [online] http://www.adm.com/news/_layouts/PressReleaseDetail.aspx?ID=589 [Accessed December 21 2015].

The Australian, (2014), *Grain assets on Japanese agenda*, June 9 2014, [online] http://www.theaustralian.com.au/business/grain-assets-on-japanese-agenda/story-e6frg8zx-1226947660517 [Accessed February 10 2015].

Baker, C., (2018), The nation building state retreats: An Australian case study in the changing role of the state, *Journal of Rural Studies*, 62, pp. 146–155.

Baumol, W., (1982), Contestable markets: An uprising in the theory of industry structure, *The American Economic Review*, 72(1), pp. 1–15.

Baumol, W., Panzar, J., and Willig, R., (1983), *Contestable markets and the theory of industry structure*, New York: Harcourt Brace Jovanich Inc.

Baumol, W., and Willig, R., (1986), *Contestability: Developments since the book*, Oxford Economic Papers, New Series, 38, pp. 9–36.

Binsted, T., (2014), *Sumitomo snaps up Emerald Grain*, *Financial Review*, [online] http://www.afr.com/business/agriculture/sumitomo-snaps-up-emerald-grain-20140206-ixuor [Accessed June 6 2017].

Blaug, M., (2001), Is competition such a good thing? Static efficiency versus dynamic efficiency, *Review of Industrial Organisation*, 19, pp. 37–48.

Broadbent Grain, (2014), *Official announcement of Broadbent Grain and CHS joint venture*, [online] https://broadbentgrain.com.au/?wpdmact=process&did=NTAuaG90bGluaw== [Accessed June 6 2017].

Brock, W., (1983), Contestable markets and the theory of industry structure: A review article, *Journal of Political Economy*, 91(6), pp. 1055–1066.

Cattle, T., and Bettles, C., (2013), *Growers welcome competition*, Farm Weekly, October 3 2013, [online] http://www.farmweekly.com.au/news/agriculture/agribusiness/general-news/growers-welcome-competition/2673569.aspx [Accessed June 5 2017].

Commonwealth of Australia, (2008a), *Parliamentary debates*. House of Representatives: official Hansard, Thursday 13 March.

Commonwealth of Australia, (2008b), *Parliamentary debates*. Senate: official Hansard, Monday 16 June.

Dasgupta, P., and Stiglitz, J., (1988), Potential competition, actual competition, and economic welfare, *European Economic Review*, 32, pp. 569–577.

Davidson, B., (2012), Contestability in human services markets, *Journal of Australian Political Economy*, 68, pp. 213–239.

Davies, J., (1986), Competition, contestability and the liner shipping industry, *Journal of Transport Economic and Policy*, 20(3), pp. 299–312.

Davies, J., and Lee, F., (1988), A post Keynesian appraisal of the contestability criterion, *Journal of Post Keynesian Economics*, 11(1), pp. 3–24.

Dennis, F., (1977), *'Competition' In the History of Economic Thought*, New York: Arno Press.

Grain Producers Australia, (2013), *Submission to the Wheat Industry Taskforce: Assessment of Wheat Export Industry Stocks Information Requirement*, [online] http://www.grainproducers.com.au/policies-and-issues/gpa-submissions/item/download/1_6522365b6877692403884c3153e5e2e3 [Accessed April 30 2015].

Grattan, M., (2008), *Split over wheat desk reform*, The Age, March 14 2008, [online] http://www.theage.com.au/news/business/split-over-wheat-desk-reform/2008/03/13/1205126110679.html [Accessed October 26 2016].

Heard, G., and Marshall, A., (2014), Infrastructure: the grain thing, The Land, September 20 2014, [online] http://www.theland.com.au/story/3381942/infrastructure-the-grain-thing/?cs=496 [Accessed October 25 2016].

Hilmer, F, Rayner, M., and Taperell, G., (1993), *National Competition Policy Review*, [online] http://ncp.ncc.gov.au/docs/National%20Competition%20Policy%20Review%20report,%20The%20Hilmer%20Report,%20August%201993.pdf [Accessed July 30 2015].

Industry Assistance Commission, (1988), *The Wheat Industry*, Report No. 411, AGPS, Canberra.

Iwata, M., (2015), *Marubeni to sow further investment in grain trader Gavilon, Wall Street Journal*, July 21 2015, [online] https://www.wsj.com/articles/marubeni-to-sow-further-investment-in-grain-trader-gavilon-1437465897 [Accessed June 6 2017].

Jasper, C., (2014), *Japanese giant moves in on Australian grains*, ABC, June 30 2014, [online] http://www.abc.net.au/news/2014-06-27/mitsubishi-olam-grains/5551698 [Accessed February 10 2015].

Jones, E., (2012), The market mechanism in reality and myth: Corporate business overlordship of small business, *Journal of Political Economy*, 68, pp. 62–85.

Marshall, A., (2015), *COFCO's grain footprint grows*, Farm Weekly, November 5 2015, [online] http://www.farmweekly.com.au/news/agriculture/cropping/general-news/cofcos-grain-footprint-grows/2747227.aspx [Accessed June 6 2017].

Murphy, S., (2006), Concentrated market power and agricultural trade, *Ecofair Trade Dialogue: Discussion Papers*, No.1 August 2006.

NSW Farmers, (2014), *Submission to Rural and Regional Affairs and Transport References Committee inquiry into Australian grain networks*, [online] http://www.aph.gov.au/DocumentStore.ashx?id=990367d0-7c89-4d83-9930-54eb96412e6e&subId=299362 [Accessed April 30 2015].

O'Keeffe, P., (2017), Creating resilience or private sector dependence? Shifting constructions of the farmer in Australian policy discourses, *Space and Polity*, 21(3), pp. 318–334.

O'Keeffe, P., and Neave, M., (2017), Experiences of wheat growers in Victoria's western Wimmera following deregulation of the Australian wheat export market, *Rural Society*, 26(1), pp. 1–18.

O'Keeffe, P., (2016a), *Responding to the productivist paradigm: Experiences of farmers in Victoria's western Wimmera*, in M. Chou (ed.) Proceedings of the Australian Sociological Association (TASA) 2016 Conference, Melbourne, Australia, 28 November–1 December 2016, pp. 263–269.

O'Keeffe, P., (2016b), Supply chain management strategies of agricultural corporations: A resource dependency approach, *Competition and Change*, 20(4), pp. 255–274.

Productivity Commission, (2005), *Review of National Competition Policy reforms: Productivity Commission Inquiry Report*, No. 33, 28 February 2005, Commonwealth of Australia, Canberra.

Queensland Bulk Terminals, (2017), *Our company*, [online] http://www.qld-bulkterminals.com/company.html [Accessed June 6 2017].

Shepherd, W., (1984), "Contestability" vs. competition, *The American Economic Review*, 74(4), pp. 572–587.

Shepherd, W., (1995), Contestability vs. competition: Once more, *Land Economics*, 71(3), pp. 299–309.

Schwartz, M., and Reynolds, R., (1983), Contestable markets: An uprising in the theory of industry structure: Comment, *The American Economic Review*, 73(3), pp. 488–490.

Sprague, J., (2014), *Bunge eyes other assets in $60m push*, Financial Review, September 22 2014, [online] http://www.afr.com/business/agriculture/bunge-eyes-other-assets-in-60m-push-20140921-jfln5 [Accessed October 25 2016].

Stretch, T., Carter, C., and Kingwell, R., (2014), *The cost of Australia's bulk grain export supply chains*, [online] http://www.aegic.org.au/media/22950/140130%20Final%20AEGIC%20Supply%20Chains%20Report.pdf [Accessed December 21 2015].

Sydney Morning Herald, (2012), *Glencore agrees to deal for Canada's Viterra*, [online] http://www.smh.com.au/business/world-business/glencore-agrees-deal-for-canadas-viterra-20120320-1vigo.html [Accessed March 17 2016].

Stiglitz, J., McFadden, D., and Peltzman, S., (1987), Technological change, sunk costs, and competition, *Brookings Papers on Economic Activity*, 3, pp. 883–947.

Tasker, S., (2012), *Japan's Mitsui buys into WA agricultural company Plum Grove*, The Australian, December 24 2012, [online] http://www.theaustralian.com.au/business/companies/japans-mitsui-buys-into-wa-agricultural-company-plum-grove/news-story/089a87ad9a397be0550e2e7e7e92c089 [Accessed June 5 2017].

Terazono, E., (2016), *Olam forms joint venture with Mitsubishi*, Financial Times, April 15 2016, [online] https://www.ft.com/content/bdd63b12-02fc-11e6-99cb-83242733f755 [Accessed June 6 2017].

Thompson, B., (2015), *CBH wants to sell terminal stake*, The West Australian, September 10 2015, [online] https://thewest.com.au/countryman/news/cbh-wants-to-sell-terminal-stake-ng-ya-128749 [Accessed June 6 2017].

Thompson, B., (2012), *Chinese giant to break WA grain grip*, The West Australian, November 10 2015, [online] https://thewest.com.au/news/australia/chinese-giant-to-break-wa-grain-grip-ng-ya-288836 [Accessed June 6 2017].

Weitzman, M., (1983), Contestable markets: An uprising in the theory of industry structure: Comment, *American Economic Review*, 73, pp. 486–487.
World Grain, (2016), *COFCO completes purchase of Noble Agri Limited*, March 3 2016, [online] http://www.worldgrain.com/articles/news_home/World_Grain_News/2016/03/COFCO_completes_purchase_of_No.aspx?ID=%7B95603DAD-6AF7-45B5-BC5B-1F4FE281EEB8%7D&cck=1 [Accessed June 5 2017].
World Grain, (2014a), *Olam, Mitsubishi partner on Australian grain business*, [online] http://www.worldgrain.com/articles/news_home/World_Grain_News/2014/06/Olam_Mitsubishi_partner_on_Aus.aspx?ID=%7bB83FC2F9-F04B-4E22-B4C7-5B46C16037D1%7d&AdKeyword&cck=1 [Accessed June 6 2017].
World Grain, (2014b), *Noble, COFCO complete agribusiness joint venture*, October 1 2014, [online] http://www.worldgrain.com/articles/news_home/World_Grain_News/2014/10/Noble_COFCO_complete_agribusin.aspx?ID=%7B85CE900D-050D-44E3-AF27-44380CA530D1%7D [Accessed June 5 2017].

CHAPTER 7

Feeding the World or Turning a Profit? How Transnational Agri-Business Firms Use Discourse to Shape Their External Environments

Introduction

Australian mainstream policy discourses around competition and markets portray firms as relatively powerless. According to this narrative, firms' market position is dependent upon their responsiveness to consumers' preferences. This construction was consistently reproduced in policy documents endorsing wheat export market deregulation (IAC 1988; BAE 1987). According to the Industries Assistance Commission (1988, p. 112), private grain traders' "involvement in trading wheat is generally motivated by commercial objectives which require that they generate profits from their trading activities." Thus, markets compel firms to respond to incentives by minimising operational costs. This increases their efficiency, and therefore, the presence of multiple firms, each responding to incentives and maximising efficiencies, is portrayed as being ultimately beneficial for growers and the broader Australian society.

Through this superior management of resources, the IAC claims that "it is quite plausible for trading organisations to make profits and, at the same time for growers to achieve higher returns...This could reflect the traders' ability to sell at prices above those obtained by the AWB in some markets and/or an ability to market wheat at lower cost than the AWB" (IAC 1988, p. 112). In addition, the IAC counteracts any negative perceptions of firms, claiming that they are no longer relevant in a modern, market-based economy (IAC 1988, pp. 111–112):

P. O'Keeffe, *Making Markets in Australian Agriculture*,
https://doi.org/10.1007/978-981-13-3519-8_7

> In the past, a major concern of growers has been that, in the absence of a central marketing authority, the market would be dominated by a few relatively large trading organisations which would be able to exercise considerable market power and suppress prices…While these concerns may have had some relevance in the 1930s, the Commission doubts their relevance to present-day economic conditions.

The reference to the 1930s is appropriate in some regards, given that unscrupulous behaviour by traders in this period motivated the formation of the Australian Wheat Board. However, the claim that a small number of multinational companies wouldn't dominate grain trade appears to overlook the dominance of the global grain industry by companies such as ADM, Bunge, Cargill, and Louis Dreyfus (often referred to as the ABCD firms), which control between 70% and 80% of the global grain trade (Clapp 2015; Murphy et al. 2012). Subsequently, as I have shown in Chap. 6, the deregulated wheat industry has essentially developed to resemble the scenario which the IAC claimed was unlikely to occur.

The concentration evident within the deregulated wheat export market should not be considered as a significant surprise. Both policy makers and transnational agri-business firms clearly indicate a preference for consolidation within the industry. According to ADM, arguing for the Australian government to approve its takeover of GrainCorp, acquisitions are essential in developing firms' competitiveness. According to ADM, there is a "clear worldwide trend towards consolidation in agribusiness, where economies of scale and global market reach are increasingly essential" (ADM 2013, p. 6). Similarly, Alberto Weisser, former Bunge CEO, contends that agri-business companies need to "be large and present in all different regions, well capitalised, have a complete portfolio. This means more consolidation, large firms with medium or even smaller ones… Bunge has been a major acquirer and will expect to continue to do so" (cited in Ahmed et al. 2014, pp. 12–13). In each of these statements, resource development in geographically diverse locations through acquisitions is considered as essential in reducing risks associated with access to inputs and also revenue volatility. Consolidation in local and global markets through acquisitions and mergers is not only normalised in this context, but it is portrayed as essential for the survival of these firms. The example of the increasingly consolidated wheat export market shows this phenomenon to be partly attributable to the merger and acquisition activities of large, transnational firms, which has removed smaller participants from the market. Policy makers' tolerance of concentrated markets is underpinned by the assumption that firms lack power in markets. This assumption, I suggest, overlooks firms' capacity to shape policy discourses, to contribute to the legitimation of consolidated agricultural markets, for the purposes of constructing a favourable operating environment.

In this chapter, I draw on speeches by prominent executives, Annual reports, media releases, shareholder announcements, and policy submissions to explore these narratives as a discursive strategy for aligning the purposes and values of the firm with those of policy makers and the broader society. This chapter contributes to literature exploring firms' capacity to influence mainstream discourses, and understandings of central themes such as climate change (Cahill 2017), corporate governance (Rooker 2015), and corporate social responsibility (Siltaoja et al. 2015; Vallentin 2015). Building upon the work of Clapp (2016), Sommerville et al. (2014), and Williams (2009), I suggest that agri-business firms' construct a discourse which portrays their expansion as being integral to global food security. Policy makers, in turn, reproduce this discourse. Thus, the business-friendly environment is not only portrayed as necessary to maximise efficiency in the wheat industry, it is integral to ensuring global food security. Firms' strategies in first developing market power and, second, exerting influence over policy discourses and policy makers show the capacity and willingness of firms to exert power over their external environment.

Legitimising Global Expansion as a Benefit to Society

The ABCD firms target investments towards firms which control resources and infrastructure, to develop their global presence. This is exemplified by Bunge's self-description as an "active acquirer of other companies" (Bunge 2015a, p. 12). This focus is part of Bunge's "Winning Global Footprint" plan, where the company seeks to "Invest in origination and distribution networks to expand grain flows" (Bunge 2014a). Expansion of business into new countries and new regions, referred to as geographical diversification, is frequently cited by large agri-business corporations as being fundamental to their global strategy (ADM 2015a, b; Louis Dreyfus 2012; Bunge 2014b). This is often portrayed as being important in securing consistent access to resources, thereby minimising revenue volatility (ADM 2015b, p. 4). From this perspective, developing a diversified global footprint is a means through which risk and uncertainty associated with resource supply can be mitigated. In the case of climatic events which may restrict the flow of resources from one region, a geographically diversified company is able to offset this reduction in resource supply through focusing on procuring resources from another region. Furthermore, as food production is seasonal, geographical expansion and diversification allow companies to ensure a consistent supply of resources, and consistent revenue streams, throughout the year. This geographical expansion is associated with the ambition to access new and developing markets (ADM 2015c, p. 4).

While geographical expansion is portrayed as essential in reducing the risks associated with resource supply and revenue volatility, the expanded reach of these companies places them in a position of significant power over global food networks. This power manifests through a presence in markets and locations across the world, which firms frequently allude to when describing their activities. For example, Juan Luciano of ADM states that:

> So ADM, as of December 31, was about $33 billion market cap[italisation]. It's certainly a global leader in origination of grain, processing of grain and ingredients. We have more than 750 facilities around the world. (ADM 2015b)

In that sense, geographic expansion, expressed through the number of facilities ADM controls, is used to portray the company's power and global reach, which, as indicated in this statement, is associated with financial strength and profitability. This omnipresence is described by Louis Dreyfus (2016):

> A glass of orange juice poured for breakfast. A bowl of rice steamed for lunch. A cup of coffee served in a neighbourhood café. Throughout the day and throughout the world, Louis Dreyfus Commodities plays a vital role in nourishing the world's population.

Cargill have also explained their pervasiveness in similar terms:

> We are the flour in your bread, the wheat in your noodles, the salt on your fries. We are the corn in your tortillas, the chocolate in your dessert, the sweetener in your soft drink. We are the oil in your salad dressing and the beef, pork or chicken you eat for dinner. We are the cotton in your clothing,

> the backing on your carpet and the fertilizer in your field". (Cargill corporate brochure 2001, cited in Murphy 2006, p. 9)

These statements serve a number of purposes. First, they convey the level of market power held by these firms, which is not immediately obvious to consumers of these products. This extends throughout the different food and clothing items and production inputs which are integral to people's everyday lives. At once, this highlights the numerous profit-making opportunities of these firms, while also suggesting that their products are inescapable. If consumers were to decide to avoid their products, they would first have difficulty differentiating which products to avoid, and second, their alternate options may be limited. In addition, these statements convey the importance of firms such as Cargill and Louis Dreyfus in feeding global populations. Above all, Cargill and Louis Dreyfus highlight their entrenched power and control within global food production and consumption.

This research has highlighted the desire of the ABCD companies to develop their ownership of integrated supply chains, particularly through investments such as acquisitions. This approach serves the dual purpose of reducing the exposure of the company to risk, while allowing the company to develop control within food systems. In addition, this enables such companies to maximise the profit that can be extracted from these chains. In that sense, control is integral, and loss of control is equated with risk. These strategies are typically framed as integral to meeting key objectives such as connecting "harvests to homes" or "farms to forks", portraying the firms as being integral to food security. Rather than producing food, the ABCD companies describe the management of supply chains as their core business; however, this is frequently constructed as serving a higher principle. Cargill CEO David MacLennan (MacLennan 2014) explains, "We move food and crops from times and places of surplus to times and places of deficit." According to Bunge (2015b, p. 15), "As we deliver on our strategies, the benefits will accrue not only to our shareholders but to society as a whole." The implication of these comments is that these firms are integral to global food networks, conveying a considerable level of power. This suggests that to disrupt the capacity of these companies to connect farms with consumers would be to disrupt global food security. In addition, this reflects an attempt to shape discourse around food policy, with government regulations restricting these strategies framed as threatening global food security.

Feeding the World

The growth of firms' global value chains and the associated opportunities for profit-making is, however, reliant upon sympathetic trade policies. Regulation inhibiting trade, such as statutory wheat marketing, is viewed by firms as a risk which potentially inhibits their ability to implement this strategy (O'Keeffe 2016). Thus, firms portray their expansion as serving a purpose to humanity, namely, ensuring global food security. This reflects the discourse of the firm, reproduced by governments, which construct firms as integral to increasing living standards. Policy makers cultivate the notion that by facilitating firms' expanded presence in social and economic life will raise consumers' living standards. Agri-business firms such as the ABCDs, on the other hand, carefully develop a public identity which suggests their core focus is upon addressing food security concerns. As Serge Schoen, then CEO of Louis Dreyfus, explains, "Our ambition is to feed and clothe people around the world by connecting farms to forks in every place where there is a need" (Louis Dreyfus 2012, p. 3). Similarly, ADM CEO Juan Luciano in his "Letter to Stockholders" in the company's 2016 Annual Report explains:

> At ADM, our purpose is to benefit society by efficiently connecting harvest to the home. We fulfil this purpose in a variety of ways through our long value chains, which extends from the farm gate to the ingredients in many products found on the consumer's plate.

Cargill Vice Chairman Paul Conway expressed similar intentions in a 2013 speech to the MIT Food Symposium, claiming that:

> Food security is what we do every day: moving foodstuffs from areas of surplus to areas of deficit...We work with NGO partners in many parts of the world to try to find long-term solutions to hunger and ways to increase agricultural productivity and incomes while simultaneously ensuring responsible use of natural resources. (Conway 2013)

These claims each share common elements. First, that the ABCD firms are integral to any viable solutions to food insecurity. Second, that their strategies of geographic expansion and vertically integrated value chains are the most efficient means of ensuring global populations have access to food. Third, that productivist farming and free trade are essential for reducing hunger. Thus, agri-business firms construct discourse around food security which places themselves as the only actors with the resources and expertise to achieve food security throughout the World (McMichael 2013; O'Keeffe 2016; Williams 2009).

ADM and Cargill have described this role as their "noble purpose" (ADM 2017; Page 2013). In particular, former Cargill Chief Executive Greg Page invokes ancient religious texts, to draw out the significance of trade to humanity:

> Trading has long been a fundamental of life. To quote Libanius, from his 'Orations III,' written in the fourth century:
>
> God did not bestow all products on all parts of the earth, but distributed his gifts over the different regions, to the end that men might cultivate a social relationship because one would have need of the help of another. And so he called commerce in to being, that all men might be able to have common enjoyment of the fruits of earth, no matter where produced.
>
> I am not sure that markets are necessarily divine, but I am convinced Libanius was on the right path.

As in the normalisation of markets in Australian policy discourses, this statement presents unrestricted trade as a central part of life across many centuries, which is the embodiment of positive, social exchange and cohesion. Page uses the extract "that all men might be able to have common enjoyment of the fruits of the Earth, no matter where produced", to highlight the divine, noble purpose of trading firms such as Cargill, which aspire to move products from "places of surplus to places of need". In addition, this quote connects commerce as a mechanism, created by God, for enabling equal distribution of the "fruits of the Earth". This portrays firms such as Cargill as being central to social cohesiveness and as the continuation of God's initial purpose for creating commerce. In this manner, ABCD firms frame their corporate strategy of expansion into new geographic regions and markets as not only integral to food security, but also the development of positive social exchange and cohesiveness. However, for firms there are two key barriers which potentially undermine their capacity to fulfil this purpose. First, the risks and uncertainties posed by government regulation, such as regulated markets and environments, as well as the imposition of taxes and tariffs, for example. Second is the real and potential competition that threatens firms' market positions. I now turn to firms' perception of these risks, before focusing on their responses to these potential threats to profitability.

Government Regulation: Risks and Responses

The preceding sections have highlighted the perceived need of agri-business to continually expand, in terms of geographic reach, and also to develop control within value chains. This expansion is often facilitated by acquisitions, and in particular, acquisitions of integrated companies. However, agri-business corporations view government regulation which impacts trade, markets, and the competitiveness of markets as potentially undermining these strategies.

ADM highlights the inherent risks that may inhibit their operation with regard to a broad range of policy areas, "including anti-trust and competition law, trade restrictions, food safety regulations, and other government regulations and mandates" (ADM 2015a, p. 12). Similarly, Bunge (2015b, p. 11) mentions, "Governmental policies affecting the agricultural industry, such as taxes, tariffs, duties, subsidies, import and export restrictions… can influence industry profitability." While these statements draw the connection between the capacity of the company to expand into new markets and regions, regulation and policy restricting this expansion is viewed as being a considerable risk to the profitability of the company.

Geographic expansion is fundamental to the strategies employed by global agri-business companies. However, as stated in the Bunge 10-K report for 2014 (Bunge 2015b, p. 11), government policies and regulations present a significant risk to trade, threatening firms' capacity to expand into new markets and regions. This underscores the concerted efforts of the ABCD companies to construct a discourse around food security which articulates that free trade, movement of capital, and global agri-business firms' expansion into new markets and regions throughout the world are of net benefit to societies. This discourse constructs regulation as a barrier which impedes firms' capacity to "feed and clothe people around the world". This is a concerted attempt to shape narratives around regulation and trade, to reduce the risk of state intervention from restricting firms' access to markets and resources, essential to their geographic expansion, and ultimately, their profitability.

In analysing this portrayal of regulation and government intervention in markets, I return to the construction of regulation in Australian policy discourses, particularly those centred on economic restructuring and competition policy. Regulation is frequently cast as a negative imposition which shackles and burdens society, preventing society from reaching its full potential. However, with regard to this discourse and the apparent need to create a "business-friendly environment", it is important to reflect upon which actors are burdened and shackled by regulation. With regard to global agri-business, market regulation, particularly through statutory marketing of commodities, inhibits firms' capacity to develop their global value chains, by expanding into new markets and regions. Thus, firms are clearly inhibited by nation states' attempts to restrict market access. This presents significant risks to profitability. In response, firms aim to construct a narrative which positions trade as essential to global food security, and themselves as benevolent actors who are best equipped to orchestrate the global trade in food and commodities.

Shaping Government Policy to Create a Business-Friendly Environment

To minimise the risks associated with government regulation and protect their power and also profitability, agri-business companies such as the ABCDs influence their environment through corporate political activity. As explained by former CEO of Cargill, Greg Page, who contended that "It is not for us to take on the roles of governments, although we should seek to inform their deliberations" (Page 2014). According to Cargill Vice Chairman Paul Conway, "we advocate for policies and practices to help ensure that the world can feed itself" (Conway 2013).

Identity is significant here. Firms construct an altruistic, benevolent identity, whereas governments are politicised, portrayed as captive to interest groups, and thus develop ill-conceived policy which may not be in society's interests. Firms, therefore, must intervene to prevent regulation which might curtail their capacity to pursue their altruistic activities. There is a striking resemblance between firms' construction of their own identity and the discursive construction of firms as efficiency-maximising actors who are governed by the needs and wants of consumers, evident in mainstream Australian policy discourses. Firms and policy makers, each of whom express an ambition to liberalise markets, produce and reproduce discourses which portray firms in this light, as powerless, beneficial actors which will raise society's living standards. Trade, between firms acting freely in open markets, rather than government intervention, is portrayed by each of these actors as essential to this aim. According to these discourses, firms, acting upon market signals communicated by consumers, will not only raise living standards but ensure the productive and efficient allocation of resources.

With regard to global food production, Page (2015) elaborates on Cargill's position, stating that food security can be reached through enabling "open trade", which therefore requires the discouragement of "political leaders worldwide from pursuing food self-sufficiency, imposing export barriers and taking other actions that inhibit food from moving freely across borders". Thus, national governments must forego attempts to develop programmes aimed at feeding their own populations, and instead rely upon Cargill, and others, to do so. Restricting market access, according to Cargill CEO David MacLennan, Vice Chairman Conway, and ADM CEO Juan Luciano, undermines food security (Luciano 2017; MacLennan 2014, 2015; Conway 2012, 2013). These arguments are communicated as directives for governments to follow, which outline the "environments that governments can provide to help provide the right incentives to farmers" (Conway 2012). The implication is that governments seeking to intervene in the activities of Cargill may risk the food security of their nation. Yet to some extent this is also the ABCD firms' vulnerability to government regulation, and highlights their perceived need to maintain some level of influence over government policy. For example, if ADM (2015a) or Bunge (2014b) considers that government regulations across areas such as competition policy and environmental regulation pose a risk to their strategy of expansion within markets and across borders, then these firms will seek to influence the development—or withdrawal—of these regulations.

Conclusion

The interventions into policy making and discourse, highlighted throughout this chapter, show firms' willingness to shape their external environments. Whereas firms use mergers and acquisitions to develop and protect their market position, firms also seek to shape policy making to create a favourable regulatory environment. The discursive construction of the firm as a benevolent, powerless efficiency maximiser, whose priorities are focused on altruistic goals such as feeding society, is evident throughout Australian mainstream agricultural, economic, and competition policy discourses in recent decades. Specifically, this construction has contributed to the deregulation of the Australian wheat export market, which has enabled firms such as Cargill, ADM, Louis Dreyfus, and Bunge to access this important market and develop their strategic position in close proximity to consumer markets throughout Asia. First, this dispels the construction of firms as passive and powerless. Clearly, firms do seek to shape their environments to create favourable outcomes. Second, this highlights firms' perceptions of risk, and the tension between risk and control.

Competition, and policy which may seek to enhance competitiveness of industry, is viewed as being a significant risk to the capacity of agri-business to implement strategies such as integration and expansion. As with the perceived need to control risks associated with access to resources, quality of product, and profit capture in value chains, intervention in policy is applied to seek some level of control over governmental decisions, and discourses informing those decisions. Thus, global agri-business firms seek to influence policy discourses, and political decision making, by producing and reproducing discourses which construct firms as benevolent organisations, which, if unrestricted by government regulation, will be free to fulfil their ambitions of enhancing societal well-being by creating employment and efficiently meeting consumers' essential needs. This is a significant source of power.

Australian policy makers' efforts to liberalise markets and industry policy are based on the presumption that firms do not possess power and do not desire to influence their external environment. In the case of the Australian wheat export market, this apparent misconception has resulted in concentrated regional markets featuring large, transnational firms. Growers, in this instance, do not have the competitive markets that they had been promised by policy makers, whose clear objective, I suggest, has been to create markets featuring large-scale participants, who, it is presumed, will operate with superior levels of productivity and efficiency and raise the performance of the wheat industry and its contribution to the national economy. Many small- to medium-sized growers have been adversely affected. However, as I argue in the previous chapter, this is not an unintended consequence of market liberalisation. This shift is designed by policy makers to reallocate resources towards the largest farmers, who have the capacity to manage relationships with transnational agri-business firms. Each of these actors are favoured by policy makers, whose aim is to enhance the allocative efficiency of the industry and shift resources towards these actors, in the presumption that a more competitive wheat industry will result from this.

As such, it is not necessarily the case that firms have deceived policy makers through the construction of their identity as the benevolent, passive, and altruistic firm. Rather, this construction helps policy makers to counter opposition about the shift towards market liberalisation in the wheat industry, for example. The firm as an efficient actor, which is integral to the state's aim of maximising efficiency in economies, industries, and markets, is portrayed as a modern conception of the firm. The firm is governed by the state, as a market steward, and by consumers, as communicators of market signals. Thus, concerns regarding collusive conduct and other market failures are outdated, as suggested by the IAC (1988). This discursive construction is developed and reproduced by policy makers and firms, to perpetuate the policy truth that policy which benefits firms, including open access to markets and minimal regulation of mergers and acquisitions, is good for society. Thus, the business-friendly environment is portrayed as being in society's best interests.

References

ADM, (2017), *The ADM way*, [online] http://www.adm.com/en-US/company/ADM_way/Pages/default.aspx [Accessed June 5 2017].

ADM, (2015a), *Form 10-K; Annual report pursuant to section 13 or 15(d) of the Securities Exchange Act of 1934, for the fiscal year ended December 31, 2013*, [online] http://www.sec.gov/Archives/edgar/data/7084/000000708414000011/adm-20131231x10k.htm [Accessed December 16 2015].

ADM, (2015b), *Archer Daniels Midland Co at Consumer Analyst Group of New York Conference*, February 17 2015, [online] http://www.adm.com/EN-US/NEWS/PRESENTATIONS/Pages/default.aspx [Accessed March 1 2016].

ADM, (2015c), *2014 Annual Report*, [online] http://www.adm.com/en-US/investors/Documents/2014-ADM-Annual-Report.pdf [Accessed March 1 2016].

ADM, (2013), *Submission to the Senate Rural and Regional Affairs and Transport References Committee: The Ownership Arrangements for Grain Handling*, [online] http://www.aph.gov.au/DocumentStore.ashx?id=ca1c7a14-741b-4d6c-a4e6-d051dd054d84&subId=5640 [Accessed December 16 2015].

Ahmed, G., Hamrick, D., and Gereffi, G., (2014), *Shifting governance structures in the wheat value chain: Implications for food security in the Middle East and North Africa*, [online] http://europa.uniroma3.it/centrorossidoria/docs/public/Ahmed_Hamrick_Gereffi_150914.pdf [Accessed April 30 2015].

Bunge Limited, (2015a), *Form 10-K; Annual report pursuant to section 13 or 15(d) of the Securities Exchange Act of 1934, for the fiscal year ended December 31, 2014,* [online] http://phx.corporate-ir.net/phoenix.zhtml?c=130024&p=irol-sec [Accessed March 7 2016].

Bunge Limited, (2015b), Unlocking Value: 2014 Annual Report, [online] http://phx.corporate-ir.net/phoenix.zhtml?c=130024&p=irol-reportsannual [Accessed March 7 2016].

Bunge Limited, (2014a), *Investor Day: Agribusiness and risk management overview,* October 12 2014, [online] http://phx.corporate-ir.net/phoenix.zhtml?c=130024&p=irol-presentations [Accessed March 7 2016].

Bunge Limited, (2014b), *Form 10-K; Annual report pursuant to section 13 or 15(d) of the Securities Exchange Act of 1934, for the fiscal year ended December 31, 2013,* [online] http://www.sec.gov/Archives/edgar/data/1144519/000104746914001600/a2218521z10-k.htm [Accessed December 16 2015].

Bureau of Agricultural Economics, (1987), *Wheat marketing and assistance,* Canberra: Australian Government Publishing Service.

Cahill, S., (2017), *Imagining alternatives in the Emerald City: The climate change discourse of transnational fossil fuel corporations,* Masters Thesis, University of Victoria.

Clapp, J., (2016), *Food, 2nd Edition,* Cambridge: Polity Press.

Clapp, J., (2015), ABCD and beyond: From grain merchants to agricultural value chain managers, *Canadian Food Studies,* 2(2), pp. 126–135.

Conway, P., (2013), *Food security: The challenge ahead,* Speech presented at the MIT Food Symposium – Cambridge, Massachusetts, November 5 2013, [online] http://www.cargill.com/news/speeches-presentations/food-security-the-challenge-ahead/index.jsp [Accessed December 16 2015].

Conway, P., (2012), *One billion farmers can't be wrong: Why prices and markets matter,* February 8 2012, [online] http://www.cargill.com/news/speeches-presentations/one-billion-farmers-prices-markets-matter/index.jsp [Accessed March 29 2016].

Industry Assistance Commission, (1988), *The Wheat Industry,* Report No. 411, AGPS, Canberra.

Louis Dreyfus, (2016), *What we do,* [online] http://www.ldcom.com/global/en/about-us/what-we-do/ [Accessed March 8 2016].

Louis Dreyfus, (2012), *2011 Annual Report,* [online] http://www.ldcommodities.com.br/Sobre_LDC/documentos/20120423_Annual%20Report_English%20Version.pdf [Accessed March 15 2016].

Luciano, J., (2017), National Grain and Feed Association industry keynote address, March 20 2017, [online] https://www.ngfa.org/wp-content/uploads/Juan-Luciano-NGFA-keynote-3-20-17.pdf [Accessed February 13 2018].

MacLennan, D., (2015), *Sustainability is the new normal*, April 21 2015, [online] http://www.cargill.com/news/sustainability-is-the-new-normal/index.jsp [Accessed March 29 2016].

MacLennan, D., (2014), *Hope for our planet: How a resilient global food system can feed a growing world*, Rabobank Duisenberg Lecture, Presented on October 11, 2014, [online] http://www.cargill.com/news/speeches-presentations/hope-for-our-planet/index.jsp [Accessed December 16 2015].

McMichael, P., (2013), Value-chain agriculture and debt relations: Contradictory outcomes, *Third World Quarterly*, 34(4), pp. 671–690.

Murphy, S., Burch, D., and Clapp, J., (2012), *Cereal Secrets: The world's largest grain traders and global agriculture*, [online] http://www.oxfam.org/sites/www.oxfam.org/files/rr-cereal-secrets-grain-traders-agriculture-30082012-en.pdf [Accessed October 28 2014].

Murphy, S., (2006), Concentrated market power and agricultural trade, *Ecofair Trade Dialogue: Discussion Papers*, No.1 August 2006.

O'Keeffe, P., (2016), Supply chain management strategies of agricultural corporations: A resource dependency approach, *Competition and Change*, 20(4), pp. 255–274.

Page, G., (2015), *Climate change and the future of food production*, October 12 2015, [online]http://www.cargill.com/news/speeches-presentations/climate-change-and-the-future-of-food-production/index.jsp [Accessed March 29 2016].

Page, G., (2014), *The price of responsibility*, Presented at the FT Commodities Summit, Switzerland, [online] http://www.cargill.com/news/speeches-presentations/the-price-of-responsibility/index.jsp [Accessed December 16 2015].

Page, G., (2013), *The price of responsibility*, [online] https://www.cargill.com/story/the-price-of-responsibility [Accessed December 18 2018].

Rooker, T., (2015), Corporate governance of governance by corporates? Testing governmentality in the context of China's national oil and petrochemical business groups, *Asia Pacific Business Review*, 21(1), pp. 60–76.

Siltaoja, M., Malin, V., and M. Pyykkönen, (2015), 'We are all responsible now': Governmentality and responsibilised subjects in corporate social responsibility, *Management Learning*, 46(4), pp. 444–460.

Sommerville, M., Essex, J., and Le Billon, P., (2014), The 'Global Food Crisis' and the geopolitics of food, *Geopolitics*, 19(2), pp. 239–265.

Vallentin, S., (2015), Governmentalities of CSR: Danish Government Policy as a reflection of political difference, *Journal of Business Ethics*, 127, pp. 33–47.

Williams, M., (2009), Feeding the World? Transnational corporations and the promotion of Genetically Modified Food, in Fuchs, D., and Clapp, J., (ed.), *Corporate power in global agrifood governance*, pp. 155–186, Cambridge, Massachusetts: MIT Press.

CHAPTER 8

Constructing a Corporate Society: Shaping Knowledge, Identities, and Values to Facilitate the Emergence of Corporate Power

Introduction

Throughout this work, I aim to understand how the Australian wheat industry came to be deregulated, and what the outcomes of this policy shift have been. Here, I draw the research articulated in the preceding chapters together, to argue that this policy shift was engineered as part of a broader restructuring of Australian society, which necessitated the reconstruction of knowledge, policy values, and identities to accord with the rationality of markets, competition, efficiency, and individualism. Governmental technologies, such as audit, performance objectives, cost-benefit analysis, and quantification, have been used to construct a governable society, according to these rationalities. This process, I suggest, has shifted power from citizens and "the public", towards consumers and corporations. Specifically, in relation to the farm sector, farmers have become increasingly marginalised as power has been transferred to transnational corporations and private capital.

In developing this work, I draw on Miller and Rose (1990, p. 8) who argue political rationalities, such as the concept of markets as apolitical mechanisms for creating efficient and productive industries, "render reality into the domain of thought". In turn, technologies of government "translate thought into the domain of reality [...] to establish 'in the world of persons and things' space and devices for acting upon those entities of which they dream and scheme" (Miller and Rose 1990, p. 8). Thus, calcu-

P. O'Keeffe, *Making Markets in Australian Agriculture*,
https://doi.org/10.1007/978-981-13-3519-8_8

lation and quantification are used to construct markets as legitimate producers of knowledge which communicates the reality of farming. This construction, I suggest, makes farming amenable to governing. In the case of the Australian wheat industry, wheat prices and costs are constructed as legitimate information. In this reality, wheat prices and costs matter, whereas other worlds have been erased. Technologies of government, including technologies of agency and performance, make this reality "operable" (Dean 1999; Miller and Rose 1990, p. 8).

Thus, this chapter consists of three sections:

1. Creating a reality of markets, firms, and consumers
2. Making society governable
3. Emerging corporate power in Australian agriculture

First, I focus on the construction of policy truths such as competition, efficiency, and markets, and how discourses have sought to create a policy environment which makes these ideas appear as logical guidelines for good policy. This, I suggest, creates a reality which is amenable to governing. In particular, I explore the construction of markets, firms, and consumers and their roles within this environment. In the next section, I turn to the restructuring of Australian agriculture, specifically, the deregulation of the wheat export market in Australia. I focus on the technologies used by government to make the rationality articulated in the first section operable. In this section, I explore the sites and actors, such as organisations and individuals, through which governing occurs. Finally, I address the emergence of firms within Australian agriculture, most particularly, the wheat export market, as powerful actors. I focus this analysis on the shifting power in the agricultural industry, towards transnational agribusiness firms and private capital, and the potential ramifications for farmers and their communities, and for food production in Australia. I conclude this chapter by reflecting on the contribution of this work and the possible areas for future governmentality research.

Creating a Reality of Markets, Firms, and Consumers

The Production and Reproduction of Ideas

In analysing the neoliberalisation of Australian agricultural policy, with a specific focus upon the wheat industry, I draw on governmentality-inspired research which highlights examples of how policy "truths" have been discursively constructed (Anderson 2010; Lockie and Higgins 2007; Pritchard 2005a, b; Tonts and Haslam-McKenzie 2005; Potter and Tilzey 2005; Oels 2005; Dixon and Hapke 2003; Higgins 2001a, b; Herbert-Cheshire 2000; Liepins and Bradshaw 1999; Liepins 1996). Adding to this body of work, I analyse how the rationality of markets, firms, and consumers, coupled with concepts such as efficiency and productivity, has been portrayed in policy documents as being the central ideas around which Australian policy, particularly agricultural policy, should be organised.

As I have highlighted in the preceding chapters, policy discourses have sought to establish competition, efficiency, and productivity as guiding principles which are unequivocally good and, ultimately, as ideas which benefit society. The logic connecting liberalised markets, competition, efficiency, and productivity with societal benefits is framed as common sense. This idea is not portrayed as a theory of how markets and firms operate; this is a self-evident reality. Competition is normalised as a part of everyday life. Furthermore, markets are described as apolitical mechanisms that simply organise transactions. I argue that this construction is used to frame markets as technical, rather than the result of political and economic interests. Conversely, policy discourses describe state intervention as politicised. At the same time, there is little reflection on whether this is actually a bad thing, given that, in many cases, including statutory wheat marketing, political intervention in markets aims to protect the collective interests of market participants such as producers. Furthermore, the liberalisation of markets, labour, finance, and capital is portrayed as inevitable. This construction is designed to weaken opposition to these policy shifts, to instead focus attention on what form this restructuring will take.

The neoliberalisation of Australian society, I suggest, has been a political project, which policy makers have implemented by shaping what matters, what has value, and how value is created, by establishing the rules, values, and norms that govern society. These discursive constructions shape policy responses to questions of how social and economic life should be organised and to delimit the potential responses to these questions, contributing to policy shifts, such as the deregulation of the wheat export market. The extension of this logic to previously non-commercial aspects of Australian life, such as education, social security, and human services, shows the pervasiveness of these ideas, but also has the potential to reshape how services are delivered in these fields, and how people relate to these services (Morley and Ablett 2016; Watts 2016; Spies-Butcher 2014, Connell 2013, Pusey 1998a, b). In this discussion, I draw on analyses of broader discourses, such as that focused on competition policy, economic policy, farming, and labour markets, which I suggest can help to reveal what policy makers have intended through the deregulation of the wheat market.

Shifting Roles and Relationships

In this section, I analyse the discursive construction of the state, firms, consumers, and workers, which I suggest is an important aspect of transferring the rationality of markets into the domain of reality. This construction helps to create a framework for making market liberalisation appear as a logical, common-sense shift that is in the interests of Australian society. I focus on the roles of these institutions and actors, their relationships, and how they have been shaped to facilitate economic restructuring in Australia, with particular attention to agriculture. As I will argue, these constructions helped operationalise policy makers' intention to create an economic environment facilitating firms' access to previously restricted areas of Australian society. This goal of corporatising parts of Australian society previously thought of as "public" has contributed, I suggest, to the emergence of corporate control in Australian agriculture.

The State and the Market

The construction of concepts such as efficiency and competition as policy truths reconfigures the role of the state. This perspective was articulated in 1986 by then Prime Minister Bob Hawke and Minister for Primary Industry John Kerin, who argued that the state's role was to create the economic conditions necessary for the most efficient and productive market participants to survive and prosper (Hawke 1986; Kerin 1986). This change in approach represents a significant shift away from the collectivist policies of the 1950s and 1960s. Rather than focusing on creating a secure and stable economic environment supporting its citizens, in more recent decades, policy discourses have constructed the idea that the state's role is to protect the efficient operation of markets. Markets, as apolitical mechanisms for organising transactions, are claimed to reward innovation and effort, and in doing so, create a fair society. This marketised society, it is assumed, will allow individuals to use their self-reliance, ingenuity, and entrepreneurialism to secure their own welfare. In this sense, risk and responsibility are devolved from the state towards individuals.

Furthermore, markets are portrayed by the state as disinterested producers of information. Transactions in markets reveal what people—as consumers—think. In doing so, markets reveal what has value and what the value of that service or object is. This policy focus upon markets also limits what is valuable to that which can be traded in markets. The pre-eminence of this construction has significant implications for policy making. For example, the policy debate around wheat export marketing deregulation in Australia was narrowed to focus on wheat prices and supply chain costs. Non-economic social and environmental consequences of this shift were externalised and portrayed as of lesser significance. In this regard, that which "matters", in a policy sense, is that which can be measured and apportioned value in markets. The question of state organisation of society then becomes focused on how to maximise production of that which is deemed valuable in markets.

This shift is part of the state's primary agricultural policy objective to facilitate the productive use of the Nation's agricultural resources. Policy makers assume increased resource productivity will contribute to good performance in economic measures such as GDP and multifactor productivity growth. These indicators, based largely on information communicated by markets, are constructed as indicators of societal well-being. These constructions legitimise policy measures designed to improve economic performance, as measured by these indicators. However, this construction relies upon a distancing of the state from society. If the state's ambition is to maximise the productive use of resources measured through economic indicators such as GDP or multifactor productivity growth, society needs to be rationalised as a collection of resources which policy makers can assemble to meet this aim. In this sense, society is made to appear less complex. However, this simplicity helps make society governable. Rather than making decisions concerning people, communities, or environments, policy makers focus on equations involving units of resources. In addition, policy change enhancing economic performance indicators, such as GDP, is justified as being in the national interest, regardless of its negative impacts upon people or communities, or the potential inequalities which figures such as GDP might conceal.

This process is evident in the case of structural adjustment in agriculture, which necessitates a distancing of policy makers from farmers and their communities and a reconfigured relationship between farmers and their land. Structural adjustment has been used in Australian agriculture as a means of increasing productive resource use through the reallocation of resources. Gray et al. (2014), the Productivity Commission (2010), and Musgrave (1990) rationalise the exits of farmers from the land as desirable in addressing this question, externalising the impact of adjustment on farmers, their families, and rural communities. This process of externalisation is made possible by conceiving of agriculture as a collection of resources, and of the state's role as creating the environment which promotes the actors which are best able to maximise the productive use of these resources. In this example, it is overall productivity that matters, justifying the exclusion of farmers from the industry.

Firms

Policy discourses construct privately owned firms as the key actors which can help the state to maximise productivity and efficiency of resource use. Whereas the state is problematised as inefficient and slow-moving, policy discourses suggest firms are compelled by the commercial disciplines of the market to maximise efficiency. In this competitive process, the firms which are least able to meet this aim will fail, whereas the most efficient firms will succeed and prosper. Firms' success is portrayed as evidence of their superior management, superior efficiencies, and their responsiveness to changing consumer needs.

The increased presence of firms is argued to benefit society through a number of ways. First, the presence of efficient firms in industries and markets will ensure resources are used productively, leading to improved economic performance. Second, firms will return efficiency gains to consumers, through lower prices. In addition, consumers will benefit from a greater range of better quality products and services. In this construction, the consumer has power, whereas firms are portrayed as lacking the capacity to exert power over markets. Even in the case of monopoly, duopoly, or oligopolies, policy discourses presume firms lack power in these scenarios, due to the threat of potential competitors entering the market and seizing their market share.

Thus, it is claimed the presence of large firms in markets is simply reflective of their superior efficiency, innovation, and responsiveness. According to this argument, efficiency is related to scale. Harper et al. (2015), among others, even use the term "efficient scale", implying that in some industries, firms require a certain size before achieving efficiency gains. As a result of this logic, policy discourses have fostered the idea that regulation restricting the growth of firms is harmful to society. In particular, policy discourses portray regulation restricting firms' capacity to develop market share as an unfair penalty upon the most successful, most efficient firms. This penalty is claimed to undermine the power of incentives for driving innovation and efficiency gains, and to reward the mediocre, less-efficient, and less-capable firms. Thus, policy discourses imply that market regulation restricting the capacity of the largest firms to develop a dominant market share, for example, injures the economy and therefore, injures consumers. In this regard, instead of a positive mechanism for protecting the interests and rights of citizens, regulation is negatively portrayed as something which undermines fairness and well-being. Within this narrative, firms are constructed as actors whose enhanced participation in society, enabled by the state, can lead to greater well-being, particularly through consumer satisfaction.

Therefore, rather than curtailing the success of firms and removing incentives for firms to grow, policy discourses claim that the state should instead aim to create a business-friendly environment. This environment, it is claimed, will stimulate investment and create more jobs, while benefitting consumers' living standards. In this regard, the interests of business become the interests of society.

The construction of firms and their power is integral to this narrative. Firms are portrayed as having limited power to influence markets, regardless of their market share or the absence of real (as opposed to potential) competition. Rather, consumers are cast as the powerful actors in this relationship. Consumers, their needs and desires, are portrayed as the key adjudicators regarding a firms' performance. Thus, firms are directed by consumers' interests and through real and potential competitors also seeking to satisfy consumers' needs and desires. The construction of this relationship between firms and consumers, moderated by markets, is presented as integral to creating a business-friendly environment. If a society consisted of consumers and firms, and if peoples' needs and desires were limited to their needs and desires as consumers, then it could be argued that firms must be liberated to meet those needs and desires.

People as Consumers

Policy discourses have sought to narrow value to that which can be created and traded in markets. In concert with this narrow construction of what has value, people are constructed as consumers. Furthermore, consumers' wants are limited to that which is deliverable in markets by firms. Thus, "what matters" in a policy sense is reduced to this relationship. As services previously delivered by the state are privatised, deregulated, and marketised, such as education, health care, telecommunications, human services, and social services, there are fewer examples where it could be argued that people are not primarily considered as consumers.

This construction fundamentally reshapes peoples' relationship to the state. Essential services are met by private firms, with a profit motive, rather than the state. In the case of service provision by the state, people relate to the service provider as a consumer, not as a taxpayer, voter, and citizen. Whereas people ostensibly have the ability to exercise power over the state in a democracy, the scope for exerting power in the customer and firm relationship is restricted. People relate to the service provider as a consumer. Their power can be exercised through choice, in the case that choice exists, and not as a taxpayer, voter, and citizen. Furthermore, in the case that choice between service providers does exist, what does that choice entail? Is there a marked difference between different private health care providers, or different private job recruitment agencies? How does the power to exercise choice between these providers measure against the power that citizens can exert over the state?

Consumers are conceived of as the ultimate beneficiaries of competition and efficiency gains stemming from market liberalisation. As policy makers frequently assert, competition benefits consumers by increasing quality and choice of goods and services, while reducing costs. This process, it is claimed, enhances living standards. However, in this context, living standards are constructed as that which can be appeased by markets and firms. As Pusey (2003) has shown, however, consumables are rarely associated with key indicators of happiness. Rather, it is relationships, friendships, family, and meaningful work that are among the different facets of life that contribute to happiness. Pusey's analysis suggests the limitations of consumers' "living standards" as an ultimate goal of efficient and competitive markets. However, the pervasiveness of the more narrowly economic understanding of living standards (and the representation of this argument as a reality) also demonstrates the capacity of discourse to shape our notions of what should contribute to happiness and to construct the idea that markets and firms are best positioned to meet these ambitions. In this sense, the construct of the consumer and its needs legitimises the extension of firms' reach into previously restricted policy areas. Thus, I suggest the "consumer" and its primacy in Australian policy making have been used as a governmental technology, as a means of devolving responsibility towards individuals in their capacities as consumers, on the one hand, and transferring responsibility from the state, towards markets and firms, on the other.

People as Farmers

The construction of people as consumers subtly changes who is intended to benefit from the policy. Mainstream economic policy discourses assume that if policy can be developed to benefit consumers, then the process leading to these ultimate consumer benefits is relatively insignificant. Whereas this narrative reduces people to consumers, peoples' role as workers and the importance of work itself are similarly marginalised. Whereas policy discourses construct consumers as powerful actors whose rights must be upheld, people as workers are portrayed differently. People's right to meaningful, secure employment and representation is constructed as less important than, and even oppositional to, consumers' interests. This portrays workers and consumers as representing two different classes of people, rather than different functions of the same person. Furthermore, workers are constructed as a resource of the "business-friendly environment", which is developed by the state to enable firms to deploy that resource most efficiently.

This approach is particularly evident in policy discourses which construct farmers as business-minded, economically rational individuals and which seek to enable these "good" farmers to prosper. Just as policy discourses construct large firms as efficiency maximisers and therefore desirable market participants, this same logic is applied to farmers. For example, policy such as the Rural Adjustment Scheme compels farmers to adopt productivist farming methods, including use of inputs and also expansion of their farms. Farmers unable to adapt to this environment risk being cast as "unviable" and exiting the industry.

As I discuss in the next section, this approach necessitates a mechanistic construction of agriculture. The value of farmers is narrowed to their ability to maximise efficiency. Farming is portrayed in one-sided terms. Rather than being constructed as a way of life entailing a deep association with rural communities, connection to land and to a family history of working on that land, farming is instead reduced to the process of converting resources to commodities. Farming is framed as a calculable operation that can be governed by addressing the simple question of what industry structure maximises efficiency. Larger farms are presumed to be the most efficient—through economies and scale, and also through the capacity to most effectively deploy technologies. Consolidated farms, operated by business people, are framed as the actors which will help the state meet the broader aims of improving the economic performance of agriculture.

In that regard, policy discourses construct consolidated farms and consolidated firms as actors which enhance industry efficiency. Policy frameworks are developed to support these actors. This development then raises the question of whether such constructions are desirable, and what the potential implications of these constructions could be, particularly for farmers and rural communities.

And What of Society?

This process of economic restructuring entails a reshaping of the roles of the state, firms, and people. In focusing social and economic policy on the policy truths of efficiency, competition, and productivity, policy discourses, I suggest, have sought to shift power away from people—as citizens, workers, and farmers, for example—towards an ill-defined population of consumers and firms. It could be generous to say that this shift has been presumed to lift Australia's overall economic performance and enhance the well-being of the Australian population. However, given that this restructuring of society has led to increased economic inequality, farmer exits, precarious employment, and stagnating wages, the weaknesses of the assumptions which have underpinned Australian policy become clear. Thus, it could be argued that this shift, at least on a structural level, expresses or has been motivated by something other than the well-being of the Australian citizenry.

Mainstream policy discourses have treated society as an externality, as Michael Pusey has argued. Workers, farmers, and regional Australia are particularly affected. In exploring the changing roles of people, the state and firms, and the relationships between these actors, I show how policy discourses have sought to create the conditions for the private ownership and control of society, particularly in relation to agriculture.

Making Society Governable

This research explores the themes outlined in the previous section by focusing on the case study of wheat export market deregulation. This policy shift entailed a substantially different approach to the collective organisation underpinning statutory wheat marketing. Liberalising wheat export markets was portrayed by policy makers, and in mainstream policy discourses, as an obvious, common-sense policy shift. The distillation of wheat marketing policy to a focus on wheat prices and supply chain costs, the construction of the AWB's purpose as being centred on the maximisation of growers' returns, and the discursive formation of the individual, self-reliant farmer who is focused principally on maximising their returns are all central elements to this. If liberalised markets provide the mechanism through which growers' wheat prices can be maximised, then this policy shift is not only in growers' best interests, but it represents the action of a responsible government. Thus, wheat export market deregulation is not the result of an ideological contest: it is simply good policy.

The previous section highlighted the discursive construction of ideas such as competition, markets, and efficiency as unequivocal truths which must be pursued in policy making if Australian society is to prosper. I now turn to the operationalisation of these concepts, through governmental technologies, whose power is drawn from their apparent neutrality and mundanity (Higgins 2002a; Rose 1999; Dean 1999; Miller and Rose 1990). I first turn to the quantification of social life, which is closely related to the technology of calculability highlighted by Rose (1991, 1993). As I suggest, quantification has permeated Australian agricultural policy making, to distil that which "matters", to quantifiable phenomena and quantitative data such as costs and benefits. This, I suggest, has helped make Australian agriculture amenable to governing. I then turn to technologies of agency, which have contributed to the reconstruction of Australian farmers as individuals whose value is encapsulated by the capacity to be productive, develop business acumen, and reposition their properties as targets for private investment. This analysis draws upon work conducted by governmentality scholars exploring the operationalisation of the self-governing individual (Laforge et al. 2017; Penny 2016; McKee 2008, 2009; Higgins et al. 2015; Lockie 2009; Lockie and Higgins 2007; Higgins and Lockie 2002; Dean 1999). My research makes the connection between the construction of the "good farmer" as one who wants the freedom and choice promised through market deregulation, and the Australian government's supposed response to what "good farmers" want: wheat export market deregulation.

Finally, in this section, I analyse the use of technologies of performance, including audit, cost-benefit analysis, benchmarking, and performance objectives, and the effect of these technologies in measuring the AWB's performance and shifting its role and purpose. Technologies of performance, as highlighted most prominently by Dean (1999), have been analysed in governmentality research (Penny 2016; Russell and Frame 2013; Russell and Thomson 2009; Rochford 2008; Leander and Munster 2007; Kurunmäki and Miller 2006; Larner 2006; O'Malley et al. 1997). My approach is slightly different, as I focus on how technologies of performance were used to govern a specific organisation, rendering its key purpose—to provide security and stability to growers through collectivised wheat marketing—as redundant.

Shaping "What Matters"

Central to the construction of policy truths such as efficiency, competition, and the market is the shaping of knowledge. By shaping what can be known, policy discourses frame how problems can be understood and the potential solutions to these problems. This process determines what can be considered as legitimate knowledge. In some ways, this process can be interpreted as a contest between the objective and the subjective, in which policy discourses concerning industry, competition, economic, and agricultural policy have sought to marginalise subjective knowledge as untrustworthy, biased, and inaccurate. On the other hand, policy discourses have actively worked to establish that "objective" data provides the only reliable means of establishing the sorts of "concrete" evidence upon which policy decisions can legitimately be made. In particular, knowledge produced by markets, such as prices and costs, is constructed as objective, unquestionable data which can be used to determine what "the truth" of a matter is. For example, in its report into the impacts of grain market deregulation, ACIL Tasman (2004) declined to consult growers. Instead, ACIL Tasman claimed that growers' buying and selling activity in markets could indicate their preferences, and thus communicate all relevant information growers could provide on the topic of grain market deregulation.

Privileging quantification is integral to this process of producing and legitimising knowledge. Quantitative data, such as that produced by markets, is constructed as reliable, objective, and neutral. In Australian policy discourses, quantification is often used interchangeably with measurement, as though the acts of quantifying and measuring are the same. However, more than simply providing the means of establishing reality, quantification enables policy makers to act upon that reality. This form of action necessitates simplifying knowledge, narrowing what can be known to that which can be quantified. This enables policy makers to view industries such as agriculture in mechanistic, reductionist terms, thereby reducing highly complex industries such as agriculture to matters which can be made solvable through calculation (Scott 1998). In turn, quantification enables policy questions to be reduced to equations which can be solved through the manipulation of numbers which are distant of the world they are claimed to represent. Policy discourses then portray these calculations as a reality. Examples include the Royal Commission into Grain Storage, Handling and Transport (1988) and its calculation of the $10 per tonne loss incurred to Australian farmers through regulation of the grain supply chain. This data simplifies the role performed by regulation, without considering the value growers and supply chain employees may have placed in these regulations. However, econometric modelling also provides policy makers with clear, unambiguous data, which can be easily communicated as truth. The accuracy of this data, in many ways, is irrelevant. What matters is that, in a context where quantitative data is constructed as an objective distillation of what reality looks like, calculations concerning the costs of regulation can be used to great political effect. In the case of the wheat export industry in Australia, these calculations were used as primary evidence of the failures of statutory wheat marketing.

Conversely, Australian agricultural policy discourses, in particular, have largely constructed qualitative data as unreliable evidence. This marginalises evidence based on lived experiences and observations. Policy discourses have eschewed the value of subjective data, indirectly restricting policy making to those who can produce quantitative data. However, this restriction also limits what can be considered in policy debates. Not all phenomena lend themselves to quantification. In particular, the social and environmental concerns arising from deregulation of Australian agricultural industries are often not easily quantified. As I show in Chap. 4, policy makers have consistently rejected the legitimacy of such concerns, claiming that they are based on data which is poorly articulated, ambiguous, not grounded in evidence, and difficult to document.

This process of constructing legitimate knowledge, through quantification in particular, is integral to shaping how the wheat industry can be understood. This narrow construction of knowledge simplifies the industry, shearing it of complexity. Importantly, that which matters, according to this construction—including prices, productivity, costs, and efficiency—is decontextualised from the complexities of the social world. Thus, the industry is made governable, by legitimising markets and separating the act of restructuring from the perceived social and environmental consequences of this structural shift. Whereas the social world is largely erased through this process, structural shifts such as farm consolidation and farmer exits are legitimised as being in the nation's interest.

Policy Values: The Economic and the Social

The discursive construction of quantitative information as the primary lens through which the issue of wheat export marketing can be understood is central to the dismantling of the AWB's single desk. Associated with this legitimation of quantitative data as knowledge, policy discourses sought to construct the industry's economic performance, measurable through growers' returns, as the only legitimate objective of wheat marketing policy. As with the marginalisation of the social through the discrediting of subjective data, policy discourses sought to portray socially-focused policy objectives, such as employment and equity, as antagonistic towards the primary ambition of efficiency. This process further narrowed the values and objectives which could legitimately be included in policy deliberations.

The IAC helped establish efficiency as the primary goal of industry and agricultural policy. Efficiency was claimed to be maximised through liberalised markets, which communicate undistorted market signals. This presumption further highlights the claimed importance of markets as producers of information. Conversely, this construction shows how regulation distorting or impeding the production of this information was constructed as contrary to the interests of the Australian society. Policy makers claimed industry policy should focus on maximising efficiency, whereas equity could be more appropriately addressed through mechanisms such as welfare. Policy makers use this argument to justify policy designed to enhance allocative efficiency, such as structural adjustment policies. Policy discourses claim equity-enhancing measures restrict the desired reallocation of resources towards the most productive resource managers, while implying that farmers who are structured out of the industry can receive equitable treatment through mechanisms such as welfare. Clearly, for farmers' sense of well-being, purpose, and self-esteem, and connection to their land, occupation, and community, the experience of being forced out of farming and into a reliance on welfare is not commensurate with equity. Policy discourses separate equity from efficiency in industry, economic, and competition policy. However, equity itself is marginalised as a legitimate policy objective. Policy discourses deride equity as reliant upon subjective interpretation. Hence, equity is constructed as a vague and value-laden concept, as opposed to the certainty communicated by markets and objectivity of information such as that expressed through prices and costs. In this regard, policy makers construct equity as being difficult to define and quantify. First, policy discourses portray equity as having different interpretations. Thus, agreement on what equity is presents a significant challenge. Conversely, it is presumed that markets communicate with surety. The market price of a commodity, such as a tonne of wheat, represents the value of this commodity, based on market demand. Second, whereas value can be measured through price, equity is less easily quantified. In this context, if equity cannot be quantified, then it cannot be based on objective data. In this manner, policy discourses subtly undermine the credibility of equity in policy making.

In addition, competition policy discourses, in particular, claim that liberalised markets are the best mechanisms for creating an equitable society. However, in making this claim, policy discourses replace equity with fairness. Furthermore, markets are constructed as mechanisms for creating fair outcomes. Yet fairness and equity are two different concepts. A system that is fair may be impartial and non-discriminatory; however, this system may not necessarily be equitable. Socially equitable policy recognises that "there are some things which people should have, that there are basic needs that should be fulfilled, that burdens and rewards should not be spread too divergently across the community" (Falk et al. 1993, p. 2). However, Australian policy discourses present fairness as related to an individuals' capacity to develop and utilise their skills, to assess and manage risks, and to respond to market signals. A fair outcome, in this regard, is one which appropriately rewards the capacity of an individual in these areas.

Markets are portrayed as fair, arising from their construction as impartial and neutral mechanisms, which produce and respond to objective data. However, markets are equipped to communicate prices based upon supply and demand of a product or service, not to understand the personal circumstances of a particular buyer and seller. Thus, what matters is the final product or service which is being presented to the market by a seller, and the ability of a consumer to pay for that product or service. In this manner, removing the concept of equity from agricultural policy is significant. Measuring farmers' value through the wheat they produce, for example, removes the significance of context. In farming, context is important.

The location and quality of land, rainfall, and other environmental factors such as frost, fire, and storms can all greatly influence a farmer's capacity to produce wheat for markets. Furthermore, structural factors may also influence farmers' participation in markets. Limited marketing knowledge, for example, may undermine farmers' market participation. However, fairness, as constructed by markets, does not consider these factors. Rather, as I discuss in the next section, a farmer's inability to develop their skills, manage risks, and respond to markets is portrayed as a result of their own failings. Thus, if farmers are not able to learn to market their wheat or to protect their farms against drought, then, policy discourses suggest, it is fair if markets do not reward farmers for anything other than the value of the wheat they are able to produce, regardless of the circumstances.

Technologies of Government

Throughout the 1980s, policy discourses around the wheat industry constructed competition, efficiency, and markets as policy truths (O'Keeffe 2016a). Farming organisations, as distinct from farmers, adopted these concepts and applied them in their vociferous critique of wheat supply chain regulation (O'Keeffe 2016a). Yet, despite the legitimation of this rationality, the wheat export market was not deregulated until 2008, following the AWB's involvement in a major international scandal. This is a curiosity. Whereas the rationality of markets was normalised in policy discourses around the wheat industry, governing regimes lacked the technologies to make deregulation possible. In particular, two key actors, farmers and the AWB, needed to be reconstructed as subjects which were amenable to governing.

Farmers

Quantification is used to construct farmers as isolated individuals who are distant from their families, communities, and land. Farmers' value is reduced to the capacity to maximise the productive use of resources. This constructs farmers as a productive unit, to be operationalised by the state. Technologies of agency provide governing regimes with the policy architecture to make this possible.

There is a substantial body of research exploring the creation of the self-reliant, self-governing farmer (e.g., Higgins 2002a, b; Higgins and Lockie 2002; Cheshire and Lawrence 2005). Policy discourses construct farmers as rational, economic individuals. Good farmers are portrayed as those who are competitive, value choice and freedom, and respond to incentives, as communicated by markets. The good farmer is not reliant upon government assistance and does not desire government support. Rather, the good farmer makes calculated decisions based upon their interpretation of market signals. Thus, whereas this farmer seeks to minimise risk, they also understand risk as being their responsibility to manage. In this regard, entrepreneurialism is constructed as moral, responsible behaviour. Policy discourses also construct good farming as an attitude, and as an adherence to a set of values, which by design reflect the broader ambitions of deregulation. Fundamentally, good farmers are constructed as interested solely in maximising their returns. The intention to maximise returns, above all else, is portrayed as common sense.

If this construction is portrayed as representative of the good farmer, then policy makers argue that policy must support these farmers. Thus, if good farmers desire the choice, freedoms, and potential to maximise returns promised by deregulation, then it can be argued that deregulation is for farmers. Conversely, policy discourses then construct regulation, not only as inhibiting the good farmer, but as supporting the less-efficient, less-responsible, and less-skilled farmers. Statutory marketing regulation is constructed as bad policy for two key reasons. First, market regulation undermines industry efficiency by preventing the reallocation of resources that may occur in a liberalised market, as farmers who are less equipped to manage the risks associated with marketing exit the industry. Second, policy discourses construct this as an issue of fairness. In this sense, regulation is unfair as it taxes the good farmers, while supporting less-capable farm managers.

The Agricultural Competitiveness White Paper is a further extension of this discursive construction. This paper portrays farms as investment targets. Rather than critique the structural changes which have undermined farmers' security, such as consolidated commodities markets, the White Paper aims to support farmers to source the necessary private capital to survive in this environment, which it constructs as a part of farming. The White Paper laments the limited private investment in Australian farming as a weakness which farmers must address to remain competitive. This framing, I suggest, transfers state responsibility for farming towards the individual capacities of farmers to attract capital.

Thus, the construction of the good farmer as a business-minded, economically rational individual who does not value their connection to land or community is an essential part of agricultural deregulation. In the past two decades in particular, Australian agricultural policy has sought to nurture this construction, which helps to make the political rationality of markets governable. Related to the narrowing of wheat marketing policy to the maximisation of growers' returns, as well as to the construction of growers as being primarily interested in maximising their returns, policy makers reconstructed the AWB's value to focus on maximising growers' returns to further the project of wheat industry deregulation.

The AWB

In 1983, the IAC identified the absence of performance objectives for the AWB as a key limitation in assessing its performance, which it presents as a key barrier to deregulation of the wheat export market. Thus, the IAC recommended establishment of performance objectives to allow policy makers—specifically, members of parliament—to assess the AWB's performance against these objectives. Performance objectives are presented as a means of making the AWB more accountable and transparent. However, I suggest that this shift intentionally narrows the AWB's purpose.

As a result, the AWB's role was narrowed to focus on maximising growers' returns while not taxing domestic consumers (IAC 1983, p. 59). This objective simplifies the AWB and makes its performance visible and legible, in the sense that its success can be easily estimated by examining the returns it is able to achieve for growers. In turn, the AWB is compelled to focus upon improving its performance according to this objective. Thus, any roles performed by the AWB which are external to its stated performance objective are diminished. In turn, the AWB also begins to perceive its value in this manner. This focus on maximising returns is thus constructed by policy makers as a reality, and accepted by the AWB as such. This acceptance then legitimises the narrowing of the AWB to focus on this objective. If the AWB argues that its value lies in its capacity to obtain premium prices through exerting market power, then policy makers can justifiably argue that its purpose should be assessed accordingly.

There is a more subtle shift associated with this use of technologies of performance. The implementation of performance objectives and measurement of performance through audit reconceptualise statutory marketing from having value for growers, towards being considered in terms of its measurable performance, which is presumed to be in consumers' interests. Critically, this performance is calculable; however, it is not comparable to the performance that would otherwise be achieved in liberalised markets. This perceived need to compare performance in regulated and unregulated markets undergirds the significance of econometric modelling in constructing an alternate reality, albeit one that is based upon questionable assumptions of firms and market behaviour. Nevertheless, this technology enabled policy makers to project the performance of liberalised markets, and compare this performance with that of the AWB as a statutory marketer. All the factors that are potentially significant, in relation to this debate around wheat export market deregulation, are narrowed to the much simpler question of which of these market structures perform better, in terms of maximising growers' returns without taxing domestic consumers. This focus marginalises arguments favouring wheat export market regulation which do not relate to growers' returns and costs borne by consumers. Subsequently, econometric modelling such as that performed by Allen Consulting (2000a) showed that liberalised markets would outperform market regulation. The limitations of this modelling, such as the assumption that deregulation of the wheat export market would result in a $5 per tonne drop in supply chain costs, escaped scrutiny, amidst the easily communicable findings of this research, which placed a monetary value on the costs of regulation upon the nation.

Consolidated Farms as Efficiency Maximisers

Allen Consulting's acknowledgement that deregulation is likely to create "winners and losers" (2000b) implies that the losers are those who cannot adapt. Farmer exits are constructed as an individual problem of poor planning and limited adaptability, rather than a structural problem. This focus on adaptability is constantly highlighted throughout policy discourses around agriculture. For example, the Productivity Commission's 2010 assessment of wheat export market deregulation and its impacts responds to growers' legitimate concerns about deregulation by asserting that growers need to adapt to this new reality. Furthermore, the 2015 Agricultural Competitiveness White Paper calls on farmers to develop "adaptive farm and business management strategies that take account of the risks they face" (Commonwealth Government 2015, p. 78). As a result, farmers' concerns around policy shifts, such as the deregulation of the wheat export market, are not received as legitimate policy criticisms. Instead, policy makers respond by implying that these changes are in the national interest and that farmers must simply adapt their practice and expectations to fit this new reality.

Policy discourses suggest that farmers who are unable to do so may incur the "short-term pain" of exiting the industry (Productivity Commission 2010, p. 86). However, this "pain" is portrayed as an insignificant problem, in relation to the presumed national benefits of restructuring. The facilitation of farmer exits involves the reallocation of farm resources from the presumably least efficient farmers to the most efficient, which is constructed as beneficial for society. This argument legitimises farm consolidation, which is assumed to be an important aspect of maximising productivity. In this sense, policy discourses prioritise the large, consolidated farms as actors which are able to maximise productive use of the nations' resources, while turning attention away from the structural problems created through this process.

Emerging Corporate Power in Australian Agriculture

Ostensibly, wheat export market deregulation was claimed to benefit wheat growers. However, as I demonstrated in the previous section, this claim is based on a construction of wheat growers as entrepreneurial, economically rational individuals. I suggest that this construction is designed to make agricultural restructuring possible, by focusing growers' identity upon economic ambitions, while decontextualising growers from their families, communities, and connection to land. Therefore, if the claim that growers benefit from deregulation is based on a politicised construction, the question of why this policy shift was implemented remains unanswered.

I suggest that it is important to view this policy shift in relation to broader policy discourses around competition policy, labour market reform, and reform in other agricultural industries. In doing so, I return to the rationality of markets as mechanisms for increasing competition and therefore maximising efficiency and productivity. I pose the question: Which market structures and which actors do policy makers consider are best equipped to maximise efficiency? I have drawn upon a collection of primarily mainstream policy discourses, produced by the Commonwealth Government, government authorities and departments, and consultancy firms employed by government. These documents focus on the wheat industry, though they also have a broader focus on agricultural, economic, and competition policy.

The common theme in these documents is that competitive markets (with competition defined as real or potential competition) maximise efficiency and productivity, which is essential in raising living standards. In addition, firms are portrayed as integral to this process. Driven by commercial disciplines, firms either operate efficiently and productively, or they succumb to more efficient and productive firms. However, there is a clear preference for large firms, believed to be the most efficient actors. Thus, policy makers are ambivalent towards consolidated markets featuring dominant, though presumably efficient, firms. Contestability theory is used by policy makers to legitimise this approach. This theory redefines competition, allowing oligopolistic and monopolistic markets to be portrayed as competitive.

This portrayal of firms, markets, and competition leads to the construction of the "business-friendly environment" as a vehicle for improving living standards. According to this argument, an environment which encourages corporate investment will attract firms and stimulate employment growth. However, the great weakness of this argument is that, in creating a business-friendly environment—one which, however, features precarious employment, underemployment, and limited employee rights and representation—the quality of life of many Australians has been affected. Nevertheless, policy makers continue to construct firms and firms' investment as integral to maximising Australia's economic performance, which, in turn, is presumed to reflect the well-being of Australian citizens.

This approach is particularly evident in the Australian government's Agricultural Competitiveness White Paper, which emphasises farmers' need to make their farms attractive to investors such as domestic superannuation funds and private equity funds. This goal is related to the Commonwealth Government's perception of agriculture as the next potential driver of economic growth, following the decline of the mining boom. To fulfil this ambition, private capital is required. Thus, farmers are important in operationalising this aim. The White Paper illustrates a range of potential investment options, each of which involve farmers ceding some level of ownership and control over their farms. The construction of private investment as integral to farming raises the possibility for farm decision making to be driven by the needs of private investors, a development which in turn has the potential to reshape agricultural production in Australia. The White Paper laments the long-term nature of farming as a potential inhibitor of investment, such as that from private equity, which has typically been used to extract short-term profits for investors. The potential land-use change emerging from this shift could exacerbate environmental problems, linked to an over-reliance on productivist farming methods (Lawrence et al. 2013; Pritchard et al. 2007). In addition, increased private investment in Australian farming could ensure that food production decision making in Australia is driven by investors, who may not make their decisions in either the national interest or farmers' interest.

This shift towards private investment is framed by the White Paper as an essential aspect of modern farming. Rather than rely upon government support, farmers instead must make their farms appealing to investors to attract the necessary capital to maintain and expand their properties. This reconstruction of farms, and of farming, has the potential to undermine farmers' autonomy, while shifting control of Australia's food production to private investors such as private equity firms or superannuation funds. This shift either is predicated on policy makers' assumption that private investors will not develop control over agriculture due to investments in farmland or reflects an ambivalence towards this potential occurring.

Returning to the concept of the "business-friendly environment", and policy makers' desire to attract investment into Australian industries, I now turn to the deregulation of the wheat export market. Proponents of this shift argued that it was in farmers' interests, and that it was all about farmers' ability to choose whom they sold their grain to. These arguments constructed a perception of competitive wheat markets featuring large numbers of buyers, with farmers holding power in their relationships with buyers. Yet the reality of wheat export market deregulation is very different. Regional grain markets, in particular, are highly concentrated. State wheat markets are controlled by dominant market actors. The dominant firms are typically those which control supply chain infrastructure. Thus, wheat markets across Australia, and wheat supply chains, are highly concentrated. For growers, this situation may not reflect the choice and freedom promised by policy makers.

The relationship between supply chain ownership and market share is particularly important. Firms recognise that, unless they develop their own supply chain infrastructure, through acquisitions or new developments, their capacity to develop a share of the wheat market will be limited. Thus, there is a significant cost associated with developing substantial market share. Consequently, these markets are not contestable. Entry is ostensibly free, but market share is dependent upon sizable investments.

Thus, entry is only genuinely available to those firms capable of making these investments. As a result, two major developments have occurred. First, firms that own infrastructure have become acquisition targets for larger firms seeking to gain entry to this market. This process has exacerbated market concentration. Second, investments in the industry, in ports and storage facilities, have only been initiated by well-capitalised, transnational value chain managers.

Policy makers could have anticipated these developments—which raises the question of whether this is an intended outcome of deregulation. Firms are constructed by policy discourses as integral to efficiency and productivity. Creating a business-friendly environment is portrayed as integral to enhancing living standards. Policy discourses assume that firms lack power and desire to influence markets and policy makers. Thus, policy makers in Australia in practice desire the involvement of large, well-capitalised firms in markets and industries. Therefore, it is difficult to reach any other conclusion than that concentrated wheat markets featuring primarily vertically integrated, transnational firms were at the very least desired by policy makers, as a scenario which would likely enhance the wheat industry's efficiency and economic performance.

Firms Influence Over Discourse, Policy, and Policy Makers

I now address the assumption that agricultural firms do not have power in markets, nor influence policy development. Responding to the risks posed by competition and regulation, firms draw together economic and political resources to shape their regulatory environment. Firms use discursive techniques to construct their strategies of expansion and growth as being integral to global food security. In this construction, firms portray food security as an issue which can only be addressed if national governments and governance agencies remove restrictions on trade and finance, and allow them to fulfil their key task of connecting "harvests to homes". In that sense, firms portray their goals and purpose as being shared by national governments and international governance agencies, yet also actively shape discourse around food industry regulation to produce a favourable policy environment.

Thus, the assumptions that firms do not have power over policy and markets are clearly false, yet are uncritically applied to Australian industry and competition policy. Policy discourses often construct firms as inherently good. Firms are described as maximising efficiency and returning gains from these efficiencies to consumers. They are constructed as responding to markets, not as influencing markets. In this regard, they are portrayed as not having power. Instead, the consumer has power. The firms' role is to satisfy consumer needs, through responding to changing preferences, as articulated by market signals.

However, this interpretation of firm behaviour is flawed. Firms use resources to develop and protect their power through markets and policy interventions. Through the inability, or unwillingness, to recognise this power, policy makers have created an environment which allows firms to develop and use their power—not just in markets, but in policy formation. Fundamentally, this environment transfers power in the industry towards agricultural corporations. Policy makers' incapacity to account for this power is a fundamental shortcoming. In addition, if we return to the basic assumptions under which the wheat export market was deregulated—that competitive pressure will compel firms to increase efficiencies and return the gains of these efficiencies to consumers—then this greater narrative is revealed as having numerous flaws. First, in the absence of competitive pressure, through actual or potential competition, it could be presumed that firms will not prioritise efficiency, and will not return any gains from efficiency to consumers, or producers. Subsequently, the economic projections of cost savings in a deregulated market appear to hold little weight. Instead, what has occurred is the shift in control from a statutory marketing authority, towards privately owned firms which retain regional monopolies, yet have minimal compulsion to either support growers or consumers. Thus, wheat export market deregulation should be viewed as a substantial failure, which has ultimately facilitated a shift in industry control towards privately owned firms.

Governmentality: The Australian Context and Beyond

What Matters? Knowledge, Construction of Truths

This research provides an interdisciplinary analysis of Australian agricultural restructuring through the case study of wheat export market deregulation. However, this work is held together by an examination of the analytics of governing. Drawing on governmentality-inspired research by Rose (1993, 1996), Miller and Rose (1990, 2008), Dean (1999), Higgins (2001a, b, 2002a, b, 2005), and Higgins and Lockie (2002) in particular,

I analyse the discursive construction of the rationality of markets, firms, and consumers. In essence, I look at how this industry was made governable, through technologies of calculation, on the one hand, and the construction of policy truths, on the other.

In relation to the former, I draw together sociology of quantification and governmentality to understand how knowledge is constructed as consisting of supposedly objective, quantitative information. Subjective information, such as the experiences and observations of farmers, in this case, is portrayed as unreliable, biased, and illegitimate information. This has the effect of erasing the social world, in favour of market-driven, quantitative information such as prices and costs. Connecting sociology of quantification and governmentality offers an opportunity to examine the governmental technologies which produce the quantitative data which is used to govern populations and analyse the construction of quantitative data as *the* legitimate source of knowledge in policy making. In addition, I draw on analysis of individuals, consumers, organisations, and corporations as offering potential new territory for governmentality research.

The Individual

This research follows numerous governmentality studies analysing the construction and operationalisation of the individual, in particular, the individual farmer. My work builds on this research by analysing the notion of the good farmer, and the relationship of this construction to smart farming and private investment. To be a smart farmer, according to more recent policy discourses, is to employ productivist methods, and to attract private investment. This extends on Higgins (2001a, b, 2002a), work in particular, which explored the farmer-as-business-person. In addition, this research analyses the construct of the good farmer, in terms of making market liberalisation possible. Whereas previous research into the "good farmer" explores this construct in relation to farm practice and appearance, I analyse the use of "the good farmer" in policy discourses around agricultural deregulation. The good farmer, in this context, is framed as one who is self-reliant and independent, but who demands choice and freedom to make their own decisions in liberalised markets. Thus, the good farmer favours market liberalisation. Using this construction, the Australian government claimed that wheat export market deregulation was in farmers' interests. In this manner, the construction of an identity—the good farmer as one who was ecumenical, and wanted the "freedom" to engage in open markets—preceded the deregulation of the Australian wheat export market. This construction was carefully crafted in policy discourses over a number of decades and helped legitimise the Australian government's eventual decision to deregulate the wheat industry. In this manner, this research explores the individualisation of farmers and how this facilitated policy change, rather than examining how policy change facilitated individualisation, as many governmentality studies have done (Higgins et al. 2015). This different approach has the capacity to be employed in other contexts, particularly in more recent examples of privatisation and marketisation, which may be occurring in Western societies where neoliberalism is now firmly entrenched.

The Consumer

In conjunction with this construction of farmers, I analyse the notion of "the consumer" as a governmental technology. Citizens are constructed as consumers who understand fairness, rather than equity. What consumers want, such as the products and services which enhance living standards, is portrayed as being deliverable in markets. My research adds to understandings of how this construct is used to justify restructuring of society, ostensibly in consumers' interests. Governmentality research is turning to the governing of the consumer, particularly in relation to ethical consumption (Powell 2018; Derkatch and Spoel 2017; Kremers and Brassett 2017; Guthman and Brown 2016; Clarke et al. 2007; Binkley 2006). However, the consumer as a site for governing extends the analysis of the entrepreneurial, self-reliant individual, as it enables society to be reconceptualised as consisting of consumers and businesses, and legitimises the transference of activity to markets. In turn, within these markets, governing is possible through the consumer and its regulated choices and actions. I suggest that there is more scope in governmentality research to analyse not only the governance of consumers but also the consumer as a technology of government, as a means of transferring activity into markets.

Organisations

This work analyses the technologies of performance used to govern through organisations, specifically the AWB. Technologies such as audit, cost-benefit analysis, and performance objectives were used to shape the AWB and facilitate the reproduction of values and practices according with the emerging neoliberal discourses through the 1980s and 1990s. Whereas governing can occur through communities, as highlighted by a substantial body of literature (Tonts and Haslam-McKenzie 2005, Herbert-Cheshire and Higgins 2004; Cheshire and Lawrence 2005; Gray and Lawrence 2001a, b), under the guise of empowerment, in this instance, governing is occurring through an organisation, with the use of performance technologies such as audit, performance objectives, and cost-benefit analysis. Examining the governance of organisations in this manner has the potential to be extended to fields of research beyond agriculture, including research into education providers and human services organisations, for example.

Corporations

There is a significant, and valuable, body of research into corporate influence in agriculture which analyses the power exerted by firms upon markets and producers, for example. However, my work departs from this research slightly, by examining the discursive shifts which have given power to firms. This builds upon research which has explored firms' capacity to influence discursive constructions of climate change (Cahill 2017), the green economy (Buseth 2017), food security (Clapp 2016; Sommerville et al. 2014; Williams 2009), corporate governance (Rooker 2015), and corporate social responsibility (Siltaoja et al. 2015; Vallentin 2015). Drawing on this work, my research challenges Australian policy discourses which position firms as being essential actors in the maximisation of consumers' living standards. According to this discourse, firms are passive efficiency maximisers, who are governed by consumers' preferences. Yet as I highlight, firms that have entered the Australian wheat industry have sought to develop their power through supply chain management strategies such as mergers, acquisitions, and joint ventures, while seeking to establish a favourable operating environment through corporate political action (O'Keeffe 2016b). In addition, as I show in Chap. 7, agricultural firms perform an active role in shaping discourse around food security and, in particular, argue that food security is most likely to be achieved if they are permitted to expand their operating model through increased trade liberalisation.

Summary

This work has analysed the case study of wheat export market deregulation, which I argue should be viewed as part of the broader restructuring of Australian society and economy. To understand the liberalisation of the wheat export market and how it was made possible, I have drawn on areas of research which at first glance might appear tacitly connected (or disconnected) from wheat export market policy. In analysing how the deregulation of the wheat export market has been made possible, I examine the construction of knowledge, values, and identities, to conform to the rationality of liberalised markets. These constructions create a reality which makes the shift from the public to the private appear as a logical, commonsense solution to the challenges facing society. I use the case study of farming and, specifically, wheat export market deregulation to show how this shift has been made possible in this context. To make this reality operable, I show how governmental technologies, such as audit, the entrepreneurial individual, cost-benefit analysis, performance objectives, econometric modelling, and the consumer, were used to act upon society, to make the shift towards liberalisation of the wheat export market happen. The construction of firms as efficiency maximisers which are relatively powerless in relation to markets and consumers is central to this shift. As a consequence, policy makers have either ignored, or failed to recognise, the capacity of firms to shape their external environments to create favourable operating conditions: a "business-friendly environment". Thus, the interests of firms have been portrayed by policy makers as essentially being the interests of the broader society. In the case of wheat export market deregulation, the liberalisation of this market has enabled transnational firms to expand their geographical footprint and extend their global value chains. On the other hand, farmers, whom policy makers claimed were the primary beneficiaries of wheat export market deregulation, contend with consolidated markets instead of choice, declining autonomy rather than individual freedom and, in some cases, feelings of disempowerment and disenfranchisement.

References

ACIL Tasman, (2004), *A review of the NCP Grain Market Reforms: NCC occasional series*, Prepared by ACIL Tasman, Canberra: Ausinfo.

Allen Consulting, (2000a), *The Wheat Marketing Act 1989 – The Economic Impact of Competitive Restrictions*, Sydney: Allen Consulting Group.

Allen Consulting, (2000b), *The Wheat Marketing Act 1989 – The Social Impact of Competitive Restrictions*, Sydney: Allen Consulting Group.

Anderson, A., (2010), *The constitution of youth: Toward a genealogy of the discourse and government of youth*, PhD Thesis, RMIT University.

Binkley, S., (2006), The perilous freedoms of consumption: Toward a theory of the conduct of consumer conduct, *Journal of Cultural Research*, 10(4), pp. 343–362.

Buseth, J., (2017), The green economy in Tanzania: From global discourses to institutionalisation, *Geoforum*, 86, pp. 42–52.

Cahill, S., (2017), Imagining alternatives in the Emerald City: The climate change discourse of transnational fossil fuel corporations, Masters Thesis, University of Victoria.

Cheshire, L., and Lawrence, G., (2005), Neoliberalism, Individualisation and Community: Regional Restructuring in Australia, *Social Identities*, 11(5), pp. 435–445.

Clapp, J., (2016), *Food, 2nd Edition*, Cambridge: Polity Press.

Clarke, N., Barnett, C., Cloke, P., and Malpass, A., (2007), Globalising the consumer: Doing politics in an ethical register, *Political Geography*, 26(3), pp. 231–249.

Commonwealth of Australia, (2015), *Agricultural Competitiveness White Paper: Stronger farmers, stronger communities*, [online] http://agwhitepaper.agriculture.gov.au/ [Accessed June 6 2016].

Connell, R., (2013), The neoliberal cascade and education: An essay on the market agenda and its consequences, *Critical Studies in Education*, 54(2), pp. 99–112.

Dean, M., (1999), *Governmentality: Power and Rule in Modern Society*, London: Sage Publications.

Derkatch, C., and Spoel, P., (2017), Public health promotion of "local food": Constituting the self-governing citizen-consumer, Health, 21(2), pp. 154–170.

Dixon, D., and Hapke, H., (2003), Cultivating discourse: The social construction of agricultural legislation, *Annals of the Association of American Geographers*, 93(1), pp. 142–164.

Falk, J., Hampton, G., Hodgkinson, A., Parker, K., and Rorris, A., (1993), *Social equity and the urban environment: Report to the Commonwealth Environment Protection Agency*, Canberra: Australian Government Publishing Service.

Gray, E., Oss-Emer, M., and Sheng, Y., (2014), *Australian agricultural productivity growth: Past reforms and future opportunities*, Australian Bureau of Agricultural and Resource Economics and Sciences, Research Report 14.2, Canberra, February.

Gray, I., and Lawrence, G., (2001a), *A future for regional Australia: Escaping global misfortune*, Cambridge: Cambridge University Press.

Gray, I., and Lawrence, G., (2001b), Neoliberalism, individualism and prospects for regional renewal, *Rural Society*, 11(3), pp. 283–298.

Guthman, J., and Brown, S., (2016), I will never eat another strawberry again: The biopolitics of consumer-citizenship in the fight against methyl iodide in California, *Agriculture and Human Values*, 33, pp. 575–585.

Harper, I., Anderson, P., McCluskey, S., and O'Bryan, M., (2015), *Competition Policy Review: Final report*, [online] http://competitionpolicyreview.gov.au/final-report/ [Accessed July 30 2015].

Hawke, B., (1986), *A message from the Prime Minister*, The Mail-Times, February 28 1986.

Herbert-Cheshire, L., and Higgins, V., (2004), From risky to responsible: expert knowledge and the governing of community-led rural development, *Journal of Rural Studies*, 20, pp. 289–302.

Herbert-Cheshire, L., (2000), Contemporary strategies for rural community development in Australia: a governmentality perspective, *Rural Society*, 16, pp. 201–215.

Higgins, V., Dibden, J., and Cocklin, C., (2015), Private agri-food governance and greenhouse gas abatement: Constructing a corporate carbon economy, *Geoforum*, 66, pp. 75–84.

Higgins, V., (2005), *Governing agriculture through the managerial capacities of farmers*, in: Higgins, V., and Lawrence, G., (eds), Agricultural Governance: Globalisation and the new politics of regulation, pp. 118–132, London: Routledge.

Higgins, V., (2002a), *Constructing reform: Economic expertise and the governing of agricultural change in Australia*, New York: Nova Science Publishers.

Higgins, V., (2002b), Self-reliant citizens and targeted populations: The case of Australian agriculture in the 1990s, *Arena Journal*, 19, pp. 161–177.

Higgins, V., (2001a), Assembling restructuring: governmentality, economic regulation and the historical emergence of the 'enterprising farmer' in Australian agricultural policy, *Review of International Political Economy*, 8(2), pp. 311–328.

Higgins, V., (2001b), Governing the boundaries of viability: Economic expertise and the production of the 'low income farm problem' in Australia, *Sociologia Ruralis*, 41(3), pp. 358–375.

Higgins, V., and Lockie, S., (2002), Re-discovering the social: neo-liberalism and hybrid practices of governing in rural natural resource management, *Journal of Rural Studies*, 18, pp. 419–428.

Industry Assistance Commission, (1983), *The Wheat Industry*, Report No. 329, AGPS, Canberra.

Kerin, R., (1986), *Poor prospects ahead*, The Mail-Times, February 28 1986.

Kremers, R., and Brassett, J., (2017), Mobile payments, social money: Everyday politics of the consumer subject, *New Political Economy*, 22(6), pp. 645–660.

Kurunmäki, L., and Miller, P., (2006), Modernising government: The calculating self, hybridisation and performance measurement, *Financial Accountability & Management*, 22(1), pp. 87–106.

Laforge, J., Anderson, C., and McLachlan, S., (2017), Governments, grassroots, and the struggle for local food systems: Containing, coopting, contesting and collaborating, *Agriculture and Human Values*, 34, pp. 663–681.

Larner, W., (2006), *Neoliberalism: Policy, ideology and governmentality*, in de Goede, M. (ed.), International political economy and poststructural politics, pp. 199–218, United Kingdom: Palgrave Macmillan.

Lawrence, G., Richards, C., and Lyons, K., (2013), Food security in Australia in an era of neoliberalism, productivism and climate change, *Journal of Rural Studies*, 29(1), pp. 30–39.

Leander, A., and Munster, R., (2007), Private security contractors in the debate about Darfur: Reflecting and reinforcing neoliberal governmentality, *International Relations*, 21(2), pp. 201–216.

Liepins, R., (1996), Reading agricultural power: Media as sites and processes in the construction of meaning, *New Zealand Geographer*, 52(2), pp. 3–10.

Liepins R, and Bradshaw, B., (1999), Neo-liberal agricultural discourse in New Zealand: Economy, culture and politics linked, *Sociologia Ruralis* 39(4), pp. 563–582.

Lockie, S., (2009), Responsibility and agency within alternative food networks: Assembling the "citizen consumer", *Agriculture and Human Values*, 26, pp. 193–201.

Lockie S, and Higgins, V., (2007), Roll-out neoliberalism and hybrid practices of regulation in Australian agri-environmental governance, *Journal of Rural Studies*, 23(1), pp. 1–11.

McKee, K., (2009), Regulating Scotland's social landlords: Localised resistance to technologies of performance management, *Housing Studies*, 24(2), pp. 155–171.

McKee, K., (2008), Transforming Scotland's public-sector housing through community ownership: The reterritorialisation of housing governance, *Space and Polity*, 12(2), pp. 183–196.

Miller, P., and Rose, N., (2008), *Governing the present: Administering economic, social and personal life*, Cambridge: Polity.

Miller, P., and Rose, N., (1990), Governing economic life, *Economy and Society*, 19(1), pp. 1–31.

Morley, C., and Ablett, P., (2016), A critical social work response to wealth and income inequality, *Social Alternatives*, 35(4), pp. 20–26.

Musgrave, W., (1990), *Rural Adjustment*, in Williams, D., (ed.), Agriculture in the Australian Economy, Sydney: Sydney University Press.

Oels, A., (2005), Rendering climate change governable: From biopower to advanced liberal government?, *Journal of Environmental Policy and Planning*, 7(3), pp. 185–207.

O'Keeffe, P., (2016a), *Towards deregulation: Shifting discourses in the Australian wheat industry*, in 2016 Asia Pacific Economic and Business History Conference Papers, Adelaide, Australia, 11–13 February 2016, pp. 1–25.

O'Keeffe, P., (2016b), Supply chain management strategies of agricultural corporations: A resource dependency approach, *Competition and Change*, 20(4), pp. 255–274.

O'Malley, P., Weir, L., and Shearing, C., (1997), Governmentality, criticism, politics, *Economy and Society*, 26(4), pp. 501–517.

Penny, J., (2016) Between coercion and consent: the politics of "Cooperative Governance" at a time of "Austerity Localism" in London, *Urban Geography*, DOI: https://doi.org/10.1080/02723638.2016.1235932.

Potter, C., and Tilzey, M., (2005), Agricultural policy discourses in the European post-Fordist transition: neoliberalism, neomercantilism and multifunctionality, *Progress in Human Geography*, 29(5), pp. 581–600.

Powell, D., (2018), Governing the (un)healthy child-consumer in the age of the childhood obesity crisis, *Sport, Education and society*, 23(4), pp. 297–310.

Pritchard, B., Burch, D., & Lawrence, G., (2007), Neither 'family' nor 'corporate' farming: Australian tomato growers as farm family entrepreneurs, *Journal of Rural Studies*, 23, pp. 75–87.

Pritchard, B., (2005a), Implementing and maintaining neoliberal agriculture in Australia, *International Journal of Sociology of Agriculture and Food*, 13(2), pp. 1–14.

Pritchard, B., (2005b), Unpacking the neoliberal approach to regional policy: A close reading of John Freebairn's 'Economic policy for rural and regional Australia', *Geographical Research*, 43(1), pp. 103–112.

Productivity Commission, (2010), *Wheat export marketing arrangements: Report no. 51*, AGPS, Canberra.

Pusey, M., (2003), *The experience of middle Australia*, Port Melbourne, Australia: Cambridge University Press.

Pusey, M., (1998a), Economic rationalism, human rights and civil society, *Australian Journal of Human Rights*, 4(2), pp. 131–152.

Pusey, M., (1998b), Women & Families: The impact of economic restructuring on women and families: Preliminary findings from the middle Australia project, *Australian Quarterly*, 70(4), pp. 18–27.

Rochford, F., (2008), Sustainable rural water delivery: balancing resource and social sustainability, *The Australasian Journal of National Resources Law and Policy*, 12(1), pp. 59–85.

Rooker, T., (2015), Corporate governance of governance by corporates? Testing governmentality in the context of China's national oil and petrochemical business groups, *Asia Pacific Business Review*, 21(1), pp. 60–76.

Rose, N., (1999), *Powers of freedom: Reclaiming political thought*, Cambridge: Cambridge University Press.

Rose, N., (1996), The death of the social? Re-figuring the territory of government, *Economy and Society,* 25(3): 327–356.

Rose, N., (1993), Government, authority and expertise in advanced liberalism, *Economy and Society,* 22(3), pp. 283–299.

Rose, N., (1991), Governing by numbers: Figuring out democracy, *Accounting Organisations and Society,* 16(7), pp. 673–692.

Royal Commission into Grain Storage, Handling and Transport, (1988), *Royal Commission into Grain Storage, Handling and Transport,* Canberra: AGPS.

Russell, S., and Frame, B., (2013), Technologies for sustainability: a governmentality perspective, *International Journal of Sustainable Development,* 16(1/2), pp. 91–106.

Russell, S., and Thomson, I., (2009), Analysing the role of sustainable development indicators in accounting for and constructing a Sustainable Scotland, *Accounting Forum,* 33, pp. 225–244.

Scott, J., (1998), *Seeing like a State: How certain schemes to improve the human condition have failed,* New Haven: Yale University Press.

Sommerville, M., Essex, J., and Le Billon, P., (2014), The 'Global Food Crisis' and the geopolitics of food, *Geopolitics,* 19(2), pp. 239–265.

Siltaoja, M., Malin, V., and M. Pyykkönen, (2015), 'We are all responsible now': Governmentality and responsibilised subjects in corporate social responsibility, *Management Learning,* 46(4), pp. 444–460.

Spies-Butcher, B., (2014), Marketisation and the dual welfare state: Neoliberalism and inequality in Australia, *The Economic and Labour Relations Review,* 25(2), pp. 185–201.

Tonts, M., and Haslam-McKenzie, F., (2005), Neoliberalism and changing regional policy in Australia, *International Planning Studies,* 10(3), pp. 183–200.

Vallentin, S., (2015), Governmentalities of CSR: Danish Government Policy as a reflection of political difference, *Journal of Business Ethics,* 127, pp. 33–47.

Watts, R., (2016), *'Running on empty': Australia's neoliberal social security system, 1988–2015,* in Mays, J., Marston, G., and Tomlinson, J., Basic income in Australia and New Zealand: Perspectives from the neoliberal frontier, pp. 69–91, Hampshire: Palgrave Macmillan.

Williams, M., (2009), Feeding the World? Transnational corporations and the promotion of Genetically Modified Food, in Fuchs, D., and Clapp, J., (ed.), *Corporate power in global agrifood governance,* pp. 155–186, Cambridge, Massachusetts: MIT Press.

Index

P. O'Keeffe, *Making Markets in Australian Agriculture*,
https://doi.org/10.1007/978-981-13-3519-8

Printed by Printforce, the Netherlands